G014

Nonmember price: $69

Member price: $55

AF335223

Handbook

on

Characterization Techniques

for the

Solid-Solution Interface

Edited by

James H. Adair
Jon A. Casey
Sridhar Venigalla

The American Ceramic Society
735 Ceramic Place
Westerville, Ohio 43081

Based in part on the proceedings of the Symposium on Characterization Techniques for the Solid-Solution Interface in Ceramic Systems, presented at the 94th Annual Meeting of the American Ceramic Society, held in Minneapolis, MN, April 12–16, 1992.

Library of Congress Cataloging-in-Publication Data

Handbook on characterization techniques for the solid-solution
    interface / edited by James H. Adair, Jon A. Casey, Sridhar
    Venigalla.
          p. cm.
      Includes index.
      ISBN 0-944904-67-X
      1. Ceramic powders. 2. Colloids. 3. Surface chemistry.
    I. Adair, James H. II. Casey, Jon A., 1954-   . III. Venigalla,
    Sridhar.
    TP815.H316   1993
    666--dc20   93-33158
                                                            CIP

Printed in the United States of America.

1 2 3 4–97 96 95 94 93

ISBN 0-944904-67-X

Preface

Colloidal chemistry and interfacial science form an integral part of ceramic powder processing science. However, the experimental techniques are often not obvious to the uninitiated nor are there any comprehensive texts available that specifically address characterization techniques relevant to colloidal systems. The objective of this volume is to present as complete as possible a collection of papers on characterization techniques for the solid-solution interface. These papers were contributed for the most part from participants at a Symposium on Characterization Techniques for the Solid-Solution Interface in Ceramic Systems, held at the 94th Annual Meeting of the American Ceramic Society in Minneapolis, MN, April 12– 16, 1992.

The intent of this volume is somewhat of a departure from the more topical papers that usually make up such proceedings. The authors in this volume were encouraged to focus on experimental techniques rather than the experimental results. In fact, authors were encouraged to discuss limitations of various solid-solution interface techniques as much as the advantages of a particular characterization tool. The volume is organized by topic with collections of papers relevant to each of the following characterization aspects: theoretical considerations in characterizing the solid-solution interface, electrokinetic methods, spectroscopic techniques, characterization of the state of dispersion of particle suspensions, and novel and innovative characterization techniques.

Any undertaking of this nature requires the combined effort of many individuals and organizations. Accordingly, we would like to express our gratitude to the American Ceramic Society, IBM Corporation, and the University of Florida Foundation for financial support in organizing the symposium. The following individuals were instrumental in reviewing and editing the papers contained herein: Sherry S. Staehle, Robert E. Chodelka, Sreeram Akunuri, Craig Habeger, Jooho Moon, Seung Beom Cho, Takayuki Tsukada, Pamela Howell, all at Materials Science and Engineering at the University of Florida, and Dr. Walter J. McGacken of Martin-Marietta Specialty Components, Inc., Largo, Florida. The success of any volume of this nature is solely dependent on the quality of the manuscripts submitted by the authors. The editors would also like to recognize the authors of each manuscript for the excellent quality of each submitted document, which made our task as straightforward as possible. Finally we would like to thank Ms. Lori Kozey and other individuals in the publications department of the American Ceramic Society for their patience and support during the review and editing process.

It is the sincere hope of the editors that the readers of this volume will find it informative and useful in their research and educational endeavors.

James H. Adair
Jon A. Casey
Sridhar Venigalla

Contents

Novel and Innovative Characterization Techniques

Theoretical Considerations in Characterizing the Solid-Solution Interface

INTERFACE CONDITIONS AND INTERPARTICLE FORCES

Roger G. Horn and Alexis Grabbe

Ceramics Division, National Institute of Standards and Technology, Gaithersburg, Maryland 20899

I. SOLID-SOLUTION INTERFACES

It is important for ceramic scientists and engineers to have an understanding of solid-liquid interfaces because there are several areas of importance to the behavior and performance of ceramic materials, for example corrosion, use of ceramics for filtration, tribology, environmental sensitivity and slow crack growth, in which the interaction between a ceramic material and a liquid environment is critical. The surface forces that we will discuss below have important roles to play in all of those areas. However, the primary reason for holding this symposium, and the primary focus of our paper, is to seek an understanding of the behavior of particles suspended in a liquid during colloidal processing of ceramics.

II. COLLOIDS

Colloidal processing provides a method of mixing and dispersing powders prior to sintering, so that powder agglomerates (which cause strength-limiting defects in the finished ceramic product) can be broken down and prevented from reforming. For colloidal processing to be effective, a slurry of powder in a liquid medium must be formed in a way that properly disperses the powders and breaks up any pre-existing agglomerates, that allows consolidation to a high solids loading, that ensures adequate strength after consolidation, and that maintains a suitable rheology throughout all of these steps.

This set of requirements calls for some rather sophisticated colloid science. The properties of colloids — stability, structure and rheology — depend on both the volume fraction of particles and on the interactions or forces acting between particles in the suspension [1,2]. These are often referred to as surface forces, because they can be considered as acting between the surfaces of the particles in close proximity.[†] Relevant forces include van der Waals attraction, electrical double-layer repulsion, steric repulsion between adsorbed polymer layers, solvation forces, and others [3,4].

Formation of a <u>stable</u> suspension (and hence dispersion of the particles) requires repulsive forces to be acting between particles. In the simplest theory of colloid stability, DLVO theory (named after its four originators [5,6]), the repulsion comes from electrostatic "double-layer" forces present between charged particles in a polar solvent such as water. Without sufficiently strong repulsive forces, the ubiquitous van der Waals attractions will cause particles to stick together whenever they collide, forming progressively larger aggregates that grow heavy enough to sediment out of suspension.

Suspensions can adopt various <u>structures</u>, for example disordered (like the molecules in a liquid) at moderate volume fractions, and ordered (like a crystal) when the volume fraction is higher or when there are long-range repulsive forces acting between particles. There can also be phase separation into liquid-like (sediment) and vapor-like (supernatant) regions, or solid-like (if the sediment is ordered) and vapor-like. Finally, under the influence of strong attractive forces, it is possible to have a <u>gel</u> or loosely-packed <u>floc</u> structure extending throughout the volume. The low packing density of such a structure is usually undesirable for ceramic processing.

The all-important <u>rheological</u> properties of colloidal systems also depend on volume fraction and on interparticle forces. Low shear-rate viscosity is high when particles are aggregated, and also when there are long-range repulsions between particles. Ordered colloids and gels can have a finite yield stress, but these structures can be broken down by the application of stress, and hence show shear-thinning behavior.

III. CHARACTERIZATION OF INTERFACES

The central theme of this paper is that the forces between particles depend to a large extent on the properties of the particle-solution interface. If we can make appropriate characterizations of that interface, we will be in a good position to calculate or at least estimate what the forces are. More importantly, we begin to understand how the interfaces might be modified in order to change the forces and hence control the behavior of the colloidal system.

In our aim of trying to show how forces are related to interface properties, we will be following a somewhat circular path, because we will in effect <u>use</u> measurements of forces as a method of characterization of interfaces. But if the force is measured directly, then according to the arguments above that is all that we

---

[†] More correctly, we should speak not of surfaces but of solid-fluid <u>interfaces</u>, but we will continue to use the terms "surfaces" and "surface forces" here.

need for understanding the colloidal behavior: we do not need any other kind of characterization of the interface in order to calculate the force.

The real benefit of the force measurements that have been made over a number of years, however, is that they have allowed us to develop a better understanding of the interface property—interparticle force relationships. This leads to better guidelines about what kind of interfacial properties to look for, and how to relate them to forces. This paper is essentially an overview of those relationships.

IV. MEASURING SURFACE FORCES

As mentioned above, one can think of making measurements of the force between particles, or model particles, as one method of characterizing the particle-solution interface. In this section we briefly describe the methods by which such force measurements can be made. The benefits of measuring forces, apart from the obvious one that they can be related directly to interparticle forces and colloidal behavior, are that such measurements are rich in information about solid-liquid interfaces, the properties of thin liquid films (including viscosity), and adsorption characteristics. There are drawbacks, however, and these will be outlined later in this section.

The history of attempts to make direct measurements of the kinds of force that act between particles in suspensions goes back more than half a century, as reviewed by Lodge [7]. Early experiments, mostly conducted in air or vacuum, were performed by the groups led by Derjaguin in Russia, Overbeek in The Netherlands, and Tabor in England. Israelachvili [8] made a major step forward by building an apparatus, based on ideas developed by Tabor, able to measure forces in a wide variety of environments, and particularly in liquids. This opened the door to a range of studies of relevance to colloid science, and most of the results alluded to in this article have been obtained with this apparatus. Similar devices have been built by various other groups, but Israelachvili's Surface Force Apparatus (SFA), now available commercially [9],[†] remains the most popular technique for measuring surface forces. For this reason, we will take a little space to describe the principles of its operation.

The Surface Force Apparatus is designed to measure forces in a <u>model</u> system, namely two smooth solids whose surfaces have cylindrical curvature. Cylinders are used because (i) thin, flat solids can easily be bent into this shape, and (ii) with the axes of the cylinders at right angles, the surfaces meet at a point, and edge effects are avoided. Crossed cylinders are (to leading order in curvature) equivalent to a sphere-flat geometry, and theories of surface forces in this geometry are well established. The cylinders have radius ~1 cm, making them rather

---

[†] Please note that specification of this or any other product does not imply its endorsement by NIST.

substantial "particles", but also making the surface forces between them large enough to be measured (the force between curved bodies being proportional to the radius of curvature). In order to measure forces at very small separations (*e.g.*, in the nanometer range), very smooth surfaces are required. Mica has been the material of choice for most SFA measurements because large sheets can be cleaved with atomically smooth surfaces.

Thin sheets of mica are glued onto cylindrical glass lenses to give them the appropriate curvature, and the lenses are mounted facing each other (see Figure 1) The distance of closest approach of the two "inner" surfaces is measured by optical interference between silver layers coated on the "outer" surfaces. Changes in surface separation can be measured with a resolution of 0.1-0.2 nm. The interference fringes also show the shape of the surfaces, and in particular a characteristic flattening when the surfaces make contact. This contact defines zero surface separation. The force between two mica surfaces is determined by mounting one of the glass lenses on a cantilever spring, and measuring the deflection of the spring. Forces as small as 100 nN can be measured in this way.

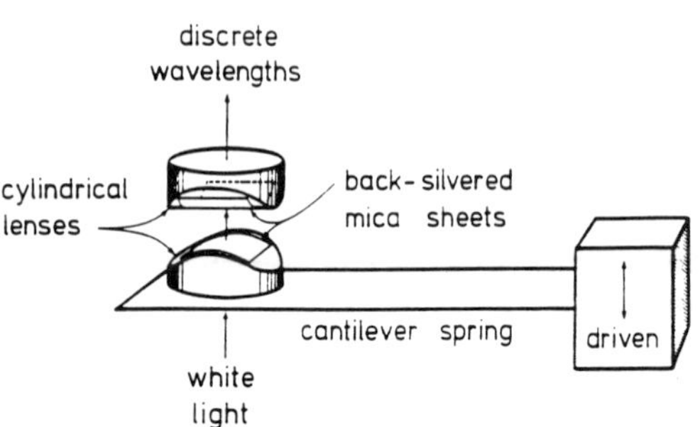

Figure 1   *Schematic picture of crossed cylinders of mica mounted in the Surface Force Apparatus. The distance between the surfaces is measured by optical interference, and force is measured from the deflection of the cantilever spring.*

The whole apparatus is a stainless steel chamber which can be filled with any liquid (or vapor) of interest, and numerous measurements have now been made with mica surfaces immersed in aqueous electrolyte solutions, surfactant solutions, polymer solutions, nonpolar liquids, polar nonaqueous liquids, and liquid mixtures. Some of these measurements will be discussed below. The mica surfaces have also been modified by adsorbed surfactant layers [10], by deposition of thin metallic [11] and metal oxide [12] films, and by chemisorption following activation by exposure to a water vapor plasma [13].

These efforts have gone some way towards broadening the range of solid-liquid interface properties that can be studied, but the technique has always been a little bit hamstrung — particularly for attempts to relate the results to real colloidal dispersions — by its reliance on mica as a model solid. More recently, however, the same SFA technique has been extended to sapphire [14] and silica [15] surfaces immersed in aqueous electrolyte solutions. The advent of new materials has several benefits: (i) sapphire and silica are themselves of direct interest to the ceramic scientist; (ii) silica lends itself to modification by chemically grafting silanes to the surface, thereby allowing control of the interfacial properties; and (iii) the possibility now exists of making force measurements between dissimilar materials. Much has been learnt about the dependence of surface forces on interfacial properties from studying one solid (mica) in combination with many different liquids. It is reasonable to expect that investigating other solid surfaces, and asymmetric combinations, will provide further clues about the mechanisms and origins of surface forces.

The main disadvantages of surface force measurements as a technique for characterizing interfaces are as follows:
(i) Even with the new materials and surface modifications now available, the range of solid materials that can be investigated remains very limited. The requirements of extreme smoothness, transparency, and thinness (to obtain optimal resolution in the optical interferometer) mean that SFA measurements cannot be made with an arbitrary material.
(ii) The experiments are very susceptible to contamination. Surfaces and liquids must be scrupulously clean to avoid any particulate or surface-active contaminants, traces of which will affect the measured forces.
(iii) The SFA is difficult to operate, being far removed from a "turn-key" apparatus. Preparation of solids and liquids, precise interferometric measurements, and data analysis all require painstaking care and a significant amount of training.

Recently a promising new technique that avoids most of these difficulties has appeared. It is based on the Atomic Force Microscope (AFM), an instrument that was developed [16] soon after the invention of the Scanning Tunneling Microscope, and which is now also commercially available [17]. As originally conceived, the AFM measures the force between a flat surface and a sharp tip. For many applications, the magnitude of this force is not the primary concern; the instrument is used to scan the tip over a surface at constant load, thereby providing a map of the surface which under favorable circumstances gives an atomic resolution image. Attempts to quantify the force as a function of tip-surface separation suffer from the difficulty of interpreting the force in this geometry. Theories are semi-quantitative at best, and the AFM has not yet proven a useful technique for improving our understanding of interparticle forces.

However, the last remark may soon be invalidated following the innovation of Ducker et al. [18], who describe a method that combines the best features of both the AFM and the SFA. They take a small silica sphere (a few μm in diameter) and glue it to the tip of a commercial AFM, then measure the force between that sphere and a flat plate. The surface forces are within the range measurable with the AFM, and the surface separation can be measured with excellent resolution (±0.3 nm). The sphere is large enough that continuum theories of surface forces are valid and applicable to the sphere-flat geometry, thereby allowing sound interpretations of the results. At the same time, the sphere is small enough that only a small surface area is involved in the interaction. The plate does not need to be particularly flat, and the likelihood of contamination becoming a problem is greatly reduced compared to the large surface areas required in the SFA. Perhaps most important of all, the solid materials no longer need to be thin and transparent, because measurement of separation is not based on optical interference. For example, forces can now be measured between materials such as silicon nitride [19].

V. THE RELATIONSHIP BETWEEN INTERFACIAL PROPERTIES AND SURFACE FORCES

In the remainder of this paper, we will discuss what is known about how interparticle forces depend on the characteristics of the particle-solution interface. Some of this knowledge is old and perhaps rather obvious; some of it is more recent, deriving primarily from direct force measurements and not yet having a strong theoretical basis, but is no less important for that.

V.1 Surface Charge

The property of a solid-solution interface that most obviously affects the force between two such interfaces is the amount of uncompensated electrical charge residing at the interface, called the <u>surface charge</u>. A solid surface typically becomes charged when immersed in a polar liquid (such as water), either because ions dissociate or dissolve from the surface, or because ions adsorb to it from the solution. Two particles of the same material acquire the same charge, and therefore repel each other. The range of the repulsion is affected by the solution properties, in particular the dielectric constant of the solvent and the concentration of electrolyte, which determine the thickness of the <u>diffuse double layer</u> in which ions of opposite sign to the surface charge (counterions) are concentrated close to the surface. The so-called <u>electrical double-layer force</u> provides an important means of dispersing particles in aqueous (and sometimes nonaqueous) solutions, provided the surface charge density is sufficiently high.

The theory of double layers and double-layer forces is well understood [1,3], and well supported by experiment [3,8,20]. Generally the double layer properties are related to electrical potential ψ. The <u>surface potential</u> ψ_0 is related to the surface charge by fundamental electrostatics, and the double-layer force is expressed in terms of surface potential. In fact there are various subtleties involved in defining exactly what one means by surface potential, subtleties that hinge on the precise distribution of charge at the interface (*e.g.*, does one consider the finite size of adsorbed ions?) [1,21]. We will not address those issues here. For the present purposes, we focus on the fact that the surface potential can be measured by various experimental methods.

The most popular methods for determining surface potential are based on electrokinetic techniques, in which the motion of charged particles in an electric

          Characterization Techniques for the Solid-Solution Interface

field is measured. The more recent development of electroacoustic methods, relying on coupling between electrokinetic and ultrasonic behavior, holds promise as a method of measuring surface potentials in dense dispersions. Both of these techniques are described in several other papers in this symposium. The surface potential so measured, called the zeta potential, is the value of the potential at the shear plane, a notional plane close to the solid-liquid interface, where the liquid obeys the no-slip hydrodynamic boundary condition. If surface potential (or charge) can be measured over a range of solution conditions such as pH, electrolyte species and electrolyte concentration, it may be possible to create a model to describe the charging mechanism, and to enable predictions of the surface potential under any specified solution conditions [21,22]. It is also possible to determine surface potential from the magnitude of the double-layer force measured in the SFA. Pashley [20] has made a comprehensive set of measurements of mica surfaces in simple electrolytes, allowing him to establish an ion-exchange model for the surface charging behavior of that material.

Metal oxide surfaces are typically positively charged at low pH, where protons attach to surface oxygens, and negatively charged at high pH, where protons are depleted from the surface, as illustrated in Figure 2. At some intermediate pH, called the point of zero charge (PZC), the surface is neutral. Behavior such as this has been established in countless studies using electrophoresis, for example. The

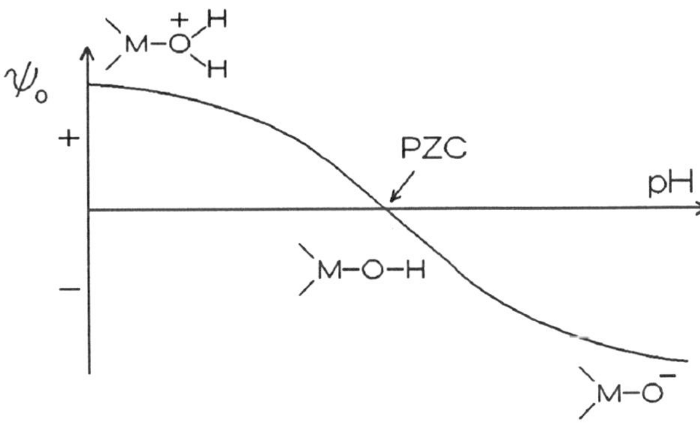

Figure 2 *Typical variation of surface potential (or surface charge) with pH for a metal oxide surface, where* M *represents the metal atom. At a particular pH, the point of zero charge (PZC), the surface is neutral.*

nature of the proton exchange interaction suggested by Fig. 2 suggests that surface charge can also be determined by potentiometric titration [1,21]. The results do not always agree with charges derived from zeta potential measurements, because titration measures a number of charge sites that may be "inside" the shear plane and not detected in the zeta potential measurements.

Clearly, if one makes a particle dispersion of a material with behavior such as that shown in Fig. 2, the electrical double-layer repulsion is sensitive to pH, and can be reduced to zero at the PZC. Thus particles can first be dispersed at a pH well above or below the PZC, then coagulated by changing pH to the PZC; this is a commonly used method of controlling colloidal stability.

With two <u>different</u> materials, the double-layer force can be attractive if the surfaces are oppositely charged. An example of this is illustrated in Figure 3, which shows SFA measurements of the force between one surface of silica and one of sapphire in water at different pH values. Sapphire has a PZC of about 9, while for silica it is at pH 2.5 [21]. Between these two pH values, sapphire is charged positively while silica is negative, and the double-layer force between them is attractive. In a mixed dispersion, the different particles would stick together, or

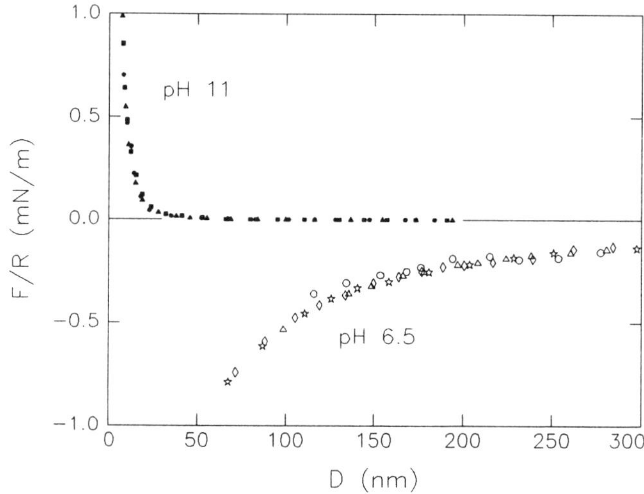

Figure 3. *Double-layer forces measured between one surface of silica and one of sapphire (basal plane) immersed in water at two different pH values. Below the PZC of sapphire (pH 9), silica is negatively charged and sapphire is positive; the double-layer force is attractive (F < 0). Above pH 9 both surfaces are negative and the force is repulsive (F > 0); the force is shorter-ranged because of the higher ionic strength at pH 11.*

<u>heterocoagulate</u>. When pH is increased above 9, however, the force becomes repulsive; all particles would be dispersed.

V.2 "Chemical Polarity"

Another vital characteristic of the solid surface is its chemical nature compared to the nature of the solvent. In particular, is the solid <u>lyophilic</u> or <u>lyophobic</u>? Lyophilic means "solvent-loving" and lyophobic means "solvent-hating"; when the solvent is water we use the terms hydrophilic and hydrophobic respectively; when it is an oil (*e.g.* a hydrocarbon liquid) we use oleophilic and oleophobic. Surfaces or chemical groups that are charged, polar or polarizable tend to be hydrophilic and oleophobic; nonpolar groups such as hydrocarbons tend to be hydrophobic and oleophilic. Here we use the term "chemical polarity" to describe this dichotomy.

The distinction between lyophilic colloids and lyophobic colloids has long been appreciated [5,6]. Lyophilic colloids, in which particles are <u>solvated</u> by the liquid, can be thermodynamically stable, so that particles disperse spontaneously and irreversibly. Lyophobic colloids are not thermodynamically stable, but can be kinetically stable (*i.e*, the particles can be dispersed with input of mechanical energy, and remain dispersed for a very long time) if appropriate repulsive interparticle forces are present; if the repulsive forces are subsequently reduced or removed, the particles coagulate.

However, there is no real theoretical basis for quantifying the lyophilicity or lyophobicity of a given solid−liquid combination. Perhaps for this reason, there sometimes seems to be less attention paid to the distinction now than in the past. For example, the classical DLVO theory of colloid stability is sometimes applied to lyophilic systems (and described as "failing") when it was never conceived for that situation. As we will elaborate below, we do have some understanding now of the correlation between lyophilicity and interparticle forces, but that understanding is based primarily on experiment, it is more qualitative than quantitative, and it has no real predictive power for solid-liquid combinations that have not previously been investigated.

The best indicator of lyophilicity is an experimental observation of whether the liquid <u>wets</u> the solid. Good wetting (a low advancing contact angle of the liquid) implies that the surface is lyophilic, and poor wetting (high contact angle) implies it is lyophobic. In some situations, however, contact angle measurements can be difficult or misleading, and the correlation with lyophilicity may not be perfect. One example is that when surface-active species are present in the liquid, the manner in which these species may adsorb to the solid and affect the surface behavior can be quite complicated, and contact angles may vary according to the exact experimental procedure. Another example, in which the contact angle is moderately high and yet the surface behaves as if it is solvated, will be given later in this section.

The clearest example of lyophilicity or <u>solvation</u> of the solid affecting forces between surfaces is the <u>hydration force</u>. Although suspected for a long time, the existence of this force was only firmly established in 1976 by the experiments of Rand, Parsegian and co-workers [23], in which a strong, short-ranged repulsive force was detected between uncharged bilayers of lecithin (a common biological membrane lipid) in a lamellar phase with water. Similar forces have since been measured with many other lipids [24]. The repulsion is attributed to the work required to remove water of hydration from the polar head group of lecithin as opposing bilayers are pushed together. The same general idea of having to remove solvent molecules from some favorable "solvation" arrangement near the solid surface is invoked to explain solvation forces in other systems.

A similar short-range (~3 nm) repulsive force was subsequently measured between mica surfaces [8,20]. In this case the force is only present above a certain critical concentration of electrolyte, dependent on the cation species. The water of hydration is actually associated with cations rather than with the mica surface itself, but a certain number of cations is bound electrostatically to the negatively-charged mica surface, and hence some water is indirectly attached to the surface. For this reason, the short-range repulsion between mica surfaces is sometimes called a "secondary" hydration force.

There is evidence that a similar secondary hydration effect occurs with alumina particles [25], except that in this case, operating at low pH, the alumina is positively charged and the hydration is mediated by anions. At neutral and alkaline pH there is no hydration force [14].

Another material for which there is clear evidence of a hydration force is silica. Direct force measurements [15,18,26,27] show that there is a short-range repulsion in aqueous electrolyte solutions. The magnitude and range (again ~3 nm) of the repulsion appear to be independent of the solution conditions, implying that the surface is inherently hydrated. Interestingly, the contact angle of water is $\approx 45°$ on the silica surface investigated in [15], so this surface appears to be less hydrophilic than mica (7°) or lecithin ($\approx 0°$), and yet is shows a hydration force of comparable strength.

In all of these cases the hydration repulsion is stronger than the van der Waals attraction, and so the surfaces are prevented from being pulled into adhesive contact. This implies that such hydrated materials should form a lyophilic, stable colloid.

As stated above, there are no theories currently capable of predicting whether a hydration force will exist with a "new" solid that has not already been investigated experimentally. Similarly, it is not clear a priori whether a repulsive force would exist between two <u>different</u> surfaces, only one of which is hydrated (in the sense that there is a hydration force between a pair of such surfaces). Recently we have addressed this question experimentally, by making measurements between the three possible combinations of mica, sapphire and silica. At least for these three materials, the answer is now clear: a hydration force is present whenever at least <u>one</u> of the surfaces is hydrated [28,29].

For example, the force measured between silica (hydrated) and sapphire (non-hydrated) always has a short-range repulsive component, even when the long-range part (shown in Fig. 3 above) is a double-layer attraction. This is shown in Figure 4. Likewise, the force between mica (hydrated above a certain cation concentration) and silica, both of which are negatively charged above pH 3, reveals a long-range double-layer repulsion, and a short-range hydration repulsion whether the cation concentration is low or high. However, between mica and sapphire, there is no hydration force at low cation concentration, and a hydration force present at high concentration, mirroring the behavior of mica by itself [29].

There is another very striking example of the importance of surface "polarity", which has also emerged from experiments on measuring surface forces. This occurs when two <u>hydrophobic</u> surfaces approach in water: an unexpectedly strong and long-range attractive force is found [10,30,31]. Experiments were conducted on mica surfaces rendered hydrophobic by adsorption of a monolayer of surfactant, with the polar head groups attaching to the mica and the hydrocarbon tails facing the water. Similar results have also been found with silica [32] and mica [13] surfaces to which hydrophobic groups have been grafted chemically.

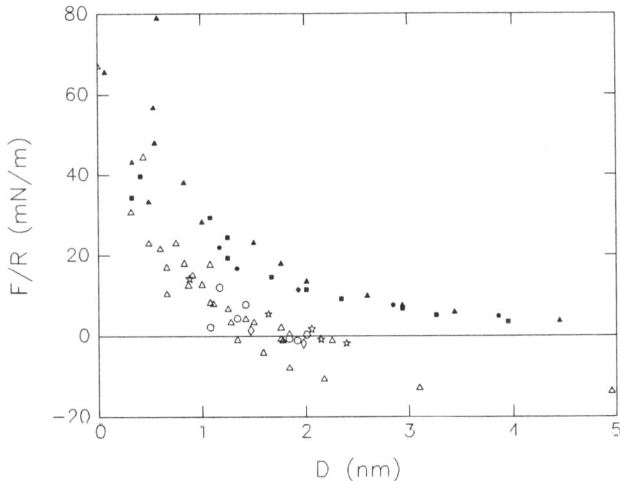

Figure 4. *The net force at short range between silica and sapphire shows a hydration repulsion superimposed on the double-layer force (see Fig. 3). At pH 6.5 (open symbols) the double-layer force is attractive, reaching about -12 mN/m at 2 nm; at pH 11 (filled symbols) the double-layer repulsion reaches about +6 mN/m at the same separation. At shorter distances the hydration repulsion dominates in both cases.*

The hydrophobic attraction has not yet been explained theoretically (at least not to universal satisfaction). Experiments show that it can be one to two orders of magnitude stronger than the van der Waals force, which implies that it is an important effect to consider if one is aiming to disperse hydrophobic particles in water. In qualitative terms, the existence of a hydrophobic attraction, even if unexpected, does not change the predicted colloidal stability: we still have a lyophobic colloid that must be stabilized by other forces. However, the magnitude of the repulsive forces required may be greater than anticipated from simple DLVO theory.

To summarize this section, the "chemical polarity" of the solid surface is an important factor in determining the interaction between solid particles. When the polarity is similar to that of the solvent, *e.g.*, a charged or polar surface in aqueous solution, the surface is more likely to be solvated, and that solvation may (though there is no guarantee of this) result in a repulsive force between surfaces that will effectively stabilize a suspension. Such a short-range force can also be beneficial in providing appropriate rheology for ceramic processing [25]. On the other hand, when the polarity is opposite to that of the solvent, *e.g.*, with hydrophobic particles in water or hydrophilic particles in nonpolar liquids, the forces that oppose dispersion of the particles may be unexpectedly strong. (See also Section V.4.)

V.3 Adsorption of surfactants and polymers

One very important means of stabilizing colloids is to coat the particles with a layer of surfactant or polymer. This method, known as steric stabilization, is commonly thought of as being a means of keeping the particles far enough apart that van der Waals forces are ineffective. In reality the situation is a little more complicated than that, since if the particle has an adsorbed layer on it, the interface should arguably be at the outer edge of that layer. Van der Waals forces still act between the two interfaces thus defined; the difference is that the strength of the van der Waals attraction (the Hamaker constant) is greatly reduced because the optical properties (on which the Hamaker constant depends [1]) of the material forming the layer are not very different from those of the surrounding liquid.

Steric stabilization adds an effective repulsive force that arises from the finite size of the adsorbate, and/or osmotic effects due to inhomogeneous distribution of polymer segments. Although the details may be subtle, a good first approximation is to think of a "hard-wall" repulsion acting at a distance given by the particle radius plus the thickness of the adsorbed layer.

The question now arises as to what are the appropriate properties of the interface that we need in order to predict steric stabilization. Do we characterize the solid-solution interface before or after adsorption of the stabilizer? As indicated above, interparticle forces are determined by the properties of the adsorbed layer: its thickness, density, and (in the case of a polymer) the distribution of polymer segments within the layer. The single most important quantity to know is the

    Characterization Techniques for the Solid-Solution Interface

amount of surfactant or polymer adsorbed per unit area of surface, and this can be measured by various means. The aforementioned SFA provides one such method (albeit not the easiest one) because the presence and thickness of adsorbed surfactant and polymer layers can be measured rather accurately as a function of concentration in the solution surrounding the surfaces. Another method is to use spectroscopy to detect adsorption at a microscopic scale, as described in other papers in this symposium.

Perhaps a more fundamental question to ask is: what determines the amount adsorbed? Could we predict it (and hence the forces) from a knowledge of properties of the interface <u>before</u> adsorption? The answer is a qualified yes: at least qualitatively but not quantitatively. Adsorption depends on the properties of three materials: the solid, the solvent, and the potential adsorbate (surfactant or polymer). Most of the behavior can be guessed from common-sense considerations. First, is the solid and/or the adsorbate charged in the solvent of interest? A cationic surfactant or polyelectrolyte will certainly adsorb to negatively charged surfaces immersed in water. Second, does the chemical nature of the adsorbate match that of the surface and/or the solvent? Since surfactants are amphiphilic molecules with both "polarities" (hydrophilic/oleophobic and hydrophobic/oleophilic) present at once, they are likely to adsorb to any surface. That is why they got their name. With polymers, one should ask whether their chemical composition resembles that of the solvent (in which case it is likely to be a "good solvent"), or not (a "poor solvent"). Adsorption is very likely to occur from a poor solvent, particularly if the particle surface has the same nature as the polymer segments. Adsorption from a good solvent is not guaranteed, although it may still occur.

Note that when polymers adsorb, there is the possibility (at low polymer concentration) of having a low adsorption density on the particle surface, which allows one polymer molecule to adsorb to more than one particle. This causes particle aggregation, leading to <u>bridging flocculation</u>, which does not aid the cause of dispersing particles.

V.4 Phase separation in mixed solvents

One final effect that can be important in a lyophobic colloid is the possibility of phase separation from a mixed solvent, whereby bridges of a second phase liquid can form between particles. This leads to very strong capillary forces that bind particles together and destabilize suspensions.

What constitutes a mixed solvent? They may be more common than we realize. For example, if oleophobic (hydrophilic) particles are suspended in a nonpolar liquid, there may be trace amounts of water present that can condense on the particle surfaces. When two particles come close together, the water forms a bridging meniscus between the particles. Even though we do not normally think of water being present in a nonpolar liquid, it does have a slight solubility (200 ppm would be typical). This means that if such a liquid is exposed to air at say 50%

humidity, the liquid becomes 50% saturated with water, *i.e.*, there would be about 100 ppm present. This may be enough to form bridges and aggregate the particles.

The important property to monitor here is again contact angle or wettability. The effect will only occur if the second phase has a low contact angle on the particle surface in the presence of the first phase — the solvent.

VI. SUMMARY

We have discussed one method of characterizing solid-solution interfaces in model systems, namely making direct surface force measurements. The technique is difficult, but it is capable of yielding a lot of information. Based largely on the knowledge gained from such experiments conducted over several years, we have endeavored in this paper to outline what we know about the relationships between properties of the interface and the interparticle forces.

REFERENCES

1. R.J. Hunter, Foundations of Colloid Science, Vols. 1 and 2. Oxford University Press, Oxford, UK, 1987 and 1989.

2. W.B. Russel, D.A. Saville and W.R. Schowalter, Colloidal Dispersions. Cambridge University Press, Cambridge, UK, 1989.

3. J.N. Israelachvili, Intermolecular and Surface Forces. Academic Press, London, UK, 2nd Ed., 1991.

4. R.G. Horn, "Surface forces and their action in ceramic materials," *J. Amer. Ceram. Soc.,* **73**, 1117-1135 (1990).

5. B.V. Derjaguin and L.D. Landau, "Theory of Stability of Highly-Charged Lyophobic Sols and Adhesion of Highly-Charged Particles in Solutions of Electrolytes," *Acta Phys. Chim. URSS* **14**, 633-652 (1941).

6. E.J.W. Verwey and J.Th.G. Overbeek, Theory of Stability of Lyophobic Colloids. Elsevier, Amsterdam, 1948.

7. K.B. Lodge, "Techniques for the Measurement of Forces between Solids," *Adv. Colloid Interface Sci.* **19**, 27-73 (1983).

8. J.N. Israelachvili and G.E. Adams, "Measurement of Forces between Two Mica Surfaces in Aqueous Electrolyte Solutions in the Range 0-100 nm," *J. Chem. Soc. Faraday Trans. I* **74**, 975-1001 (1978).

9. Anutech Pty. Ltd., GPO Box 4, Canberra, ACT 2600, Australia.

10. J.N. Israelachvili and R.M. Pashley, "Measurement of the Hydrophobic Interaction between Two Hydrophobic Surfaces in Aqueous Electrolyte Solutions," *J. Colloid Interface Sci.* **98**, 500-514 (1984).

11. C.P. Smith, M. Maeda, L. Atanasoska and H.S. White, "Ultrathin platinum films on mica and the measurement of forces at the platinum/water interface," *J. Phys. Chem.* **92**, 199-206 (1988).

12. S.J. Hirz, A.M. Homola, G. Hadziioannou and C.W. Frank, "Effect of substrate on shearing properties of ultrathin polymer films," *Langmuir* **8**, 328-333 (1992).

          Characterization Techniques for the Solid-Solution Interface

13.   J.L. Parker, D.L. Cho and P.M. Claesson, "Plasma Modification of Mica: Forces between Fluorocarbon Surfaces in Water and a Nonpolar Liquid," *J. Phys. Chem.* **93**, 6121-6125 (1989).

14.   R.G. Horn, D.R. Clarke and M.T. Clarkson, "Direct Measurement of Surface Forces between Sapphire Crystals in Aqueous Solutions," *J. Mater. Res.* **3**, 413-416 (1988).

15.   R.G. Horn, D.T. Smith and W. Haller, "Surface Forces and Viscosity of Water Measured between Silica Sheets," *Chem. Phys. Letters* **162**, 404-408 (1989).

16.   G. Binnig, C.F. Quate and C. Gerber, "Atomic Force Microscope," *Phys. Rev. Lett.* **56**, 930-933 (1986).

17.   For example, Digital Instruments, Inc., 6780 Cortona Drive, Santa Barbara, California 93117; Park Scientific Instruments, 476 Ellis Street, Mountain view, California 94043; Omicron Associates, 1738 N. Highland Road, Pittsburgh, Pennsylvania 15241.

18.   W.A. Ducker, T.J. Senden and R.M. Pashley, "Direct measurement of colloidal forces using an atomic force microscope," *Nature* **353**, 239-241 (1991).

19.   T.J. Senden and C.J. Drummond, "Surface forces between silicon nitride surfaces in aqueous electrolyte solutions: measurement with an atomic force microscope" Paper presented at the 203rd American Chemical Society National Meeting, San Francisco, April 5-10, 1992.

20.   R.M. Pashley, "DLVO and Hydration Forces Between Mica Surfaces in Li^+, Na^+, K^+, and Cs^+ Electrolyte Solutions: A Correlation of Double-Layer and Hydration Forces with Surface Cation Exchange Properties," *J. Colloid Interface Sci.* **83**, 531-546 (1981).

21.   R.O. James, "Characterization of colloids in aqueous systems;" pp 349-410 in Advances in Ceramics, Vol. 21: Ceramic Powder Science. Edited by G.L. Messing, K.S. Mazdiyasni, J.W. McCauley and R.A. Haber, American Ceramic Society, Westerville, Ohio, 1987.

22.   T.W. Healy and L.R. White, "Ionizable surface group models of aqueous interfaces," *Adv. Colloid Interface Sci.* **9**, 303-345 (1978).

23.   D.M. LeNeveu, R.P. Rand and V.A. Parsegian, "Measurement of Forces between Lecithin Bilayers," *Nature* **259**, 601-603 (1976).

24.   R.P. Rand and V.A. Parsegian, "Hydration forces between phospholipid bilayers," *Biochim. Biophys. Acta* **988** 351-376 (1989).

25.   B.V. Velamakanni, J.C. Chang, F.F. Lange and D.S. Pearson, "New method for efficient colloidal particle packing via modulation of repulsive lubricating hydration force," *Langmuir* **6**, 1323-1325 (1990).

26.   Ya.I. Rabinovich, B.V. Derjaguin and N.V. Churaev, "Direct Measurements of Long-Range Surface Forces in Gas and Liquid Media," *Adv. Colloid Interface Sci.* **16**, 63-78 (1982).

27.   G. Peschel, P. Belouschek, M.M. Müller, M.R. Müller and R. König, "The Interaction of Solid Surfaces in Aqueous Systems," *Colloid Polymer Sci.* **260**, 444-451 (1982).

28.   A. Grabbe and R.G. Horn, "Direct measurements of double-layer and hydration forces between oxide surfaces," to be published in Proceedings of NATO Advanced Research Workshop on Clay Swelling and Expansive Soils. Editors P. Baveye and M.B. McBride, Kluwer Scientific (in press).

29.   R.G. Horn and A. Grabbe, "Double-layer and hydration forces between dissimilar materials," in preparation.

30.   R.M. Pashley, P.M. McGuiggan, B.W. Ninham and D.F. Evans, "Attractive Forces between Uncharged Hydrophobic Surfaces: Direct Measurements in Aqueous Solution," *Science* **229**, 1088-1089 (1985).

31.   P.M. Claesson and H.K. Christenson, "Very Long-Range Attractive Forces between Uncharged Hydrocarbon and Fluorocarbon Surfaces in Water," *J. Phys. Chem.* **92**, 1650-1655 (1988).

32.   Ya.I. Rabinovich and B.V. Derjaguin, "Interaction of Hydrophobized Filaments in Aqueous Electrolyte Solutions," *Colloids Surfaces* **30**, 243-251 (1988).

CHARACTERIZATION TECHNIQUES AND PROPERTIES OF NONAQUEOUS MAGNETIC COLLOIDS

V. J. Novotny, IBM Research Division, Almaden Research Center, 650 Harry Road, San Jose, California 95120-6099

Properties of magnetic colloids are often dominated by magnetic attractive dipole-dipole forces. Steric stabilization with surfactants is used to reduce these attractive interactions so that rheological properties are improved and thin, smooth coatings without defects can be produced for particulate magnetic recording applications. This study concentrates on particle surface, solution, and colloid characterization of a system consisting of iron oxide particles, phosphate ester surfactant and binder in cyclohexanone solvent. Particle composition is examined by x-ray photoelectron spectroscopy (xps) and acid/base character of particle surface by surface calorimetry. Ionic mobilities and concentrations are investigated by electrical transients and particle mobilities by video electrophoresis. Adsorption isotherms of surfactant are determined by plasma emission spectroscopy of supernatant and xps of particle surface. Surfactant-particle interaction energies are also measured calorimetrically. Based on this extensive characterization, electrostatic, van der Waals, steric and magnetic interaction energies are calculated. Magnetic attractions predominate and the colloid is unstable. Smaller particle size and better steric stabilization are required to improve these colloids.

INTRODUCTION

Particulate magnetic recording media which are used to store digital or analog data have to satisfy a number of requirements. In order to achieve high storage densities, the particulate coating has to be thin, usually 100-1000 nm. It has to have high pigment volume concentration of magnetic particles, usually γ-Fe_2O_3, to provide high signal during readout. The coating has to be planar, without defects and with minimum density of asperities, so that the magnetic head attached to the slider can fly above the particulate media surface at separations of 100-500 nm. The coating has to be also wear resistant in order to sustain head-disk interactions and contact start-stops in many systems. These particulate media requirements translate quite directly to colloidal requirements for magnetic colloids which are used to prepare the coating.

The magnetic colloid should be well dispersed, without large agglomerates and preferably stable. Magnetic interactions between particles are generally very strong and therefore steric and electrostatic interactions have to be relied on to impart to the colloid at least partial stability. In order to improve the colloid, its components which include particles, surfactant, binder and suspending liquid have to be analyzed, their interactions measured and colloidal interactions understood at least on an elementary level. Colloidal interaction energies are schematically sketched in Fig. 1 together with the basic parameters characterizing the interactions between colloidal particles.

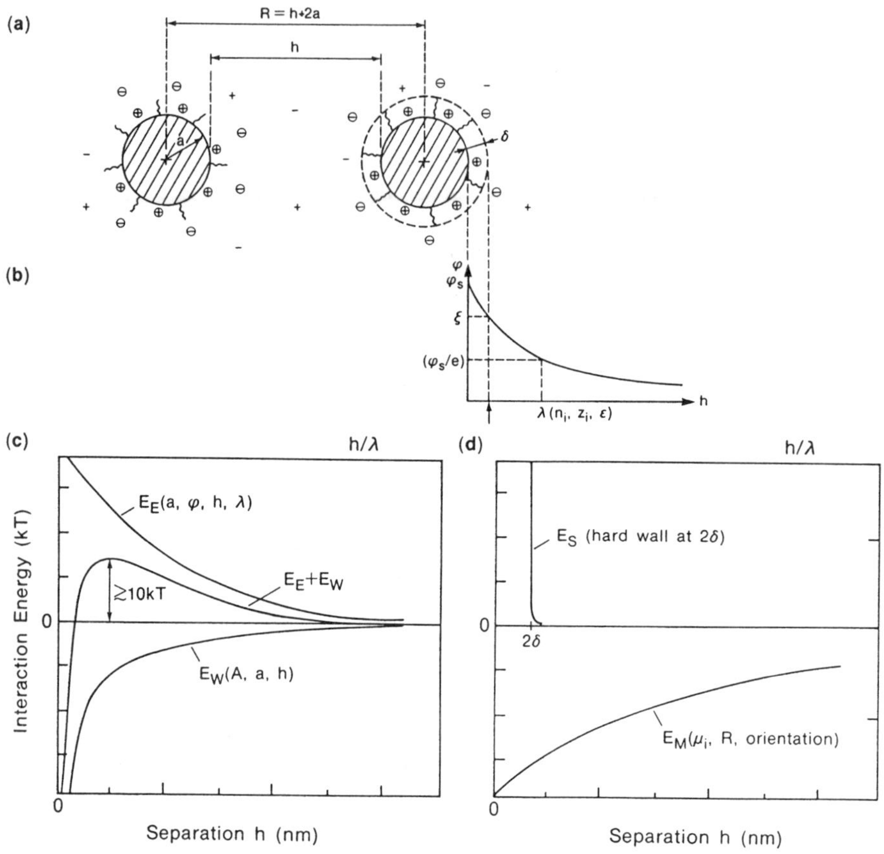

Figure 1: Schematics of (a) two charged colloidal particles with their electrical double layers (b) their electrostatic potential, φ, vs distance, h, (c) electrostatic, E_E, and van der Waals, E_W, and (d) steric, E_S, and magnetic, E_M, interaction energies.

Electrostatic interaction energy of two charged particles surrounded by their electrical double layers depends on the particle radius, a, separation of particles, h, their electrostatic potential, φ, and the characteristic width of the double layer, λ, called the Debye length.[1] This length, in turn, is a function of the concentration of ionic species, n_i, their valency, z_i, and the dielectric constant, ε, of the media:

        Characterization Techniques for the Solid-Solution Interface

$$\lambda = \left[\varepsilon\varepsilon_0 kT/(e^2 \sum_i n_i z_i^2) \right]^{1/2} \tag{1}$$

where ε_0 is the vacuum permitivity, k is the Boltzman's constant, T is the temperature, e is the unit electronic charge and the summation i is over all ionic species. For colloids with high or intermediate electrical conductivity (of the order of 10^{-9} to 10^{-10} S/m and higher), the Debye length, λ, is usually much smaller than the particle radius, a, and consequently the particle charge is

$$q = 4\pi\eta\mu a \tag{2}$$

where η is the viscosity of the suspending liquid and μ is the particle mobility. The electrostatic potential at the plane of shear is given by

$$\xi = \eta\mu/(\varepsilon\varepsilon_0) \tag{3}$$

and the electrostatic interaction energy is[1]

$$E_E = 2\pi\varepsilon\varepsilon_0\varphi_s^2 a \ln[1 + \exp(-h/\lambda)] \tag{4}$$

where φ_s is the surface potential. The van der Waals interactions due to induced dipole interactions are given by

$$E_W = -\frac{A}{6}\left(\frac{2}{s^2 - 4} + \frac{2}{s^2} + \ln\frac{s^2 - 4}{s^2} \right). \tag{5}$$

$s = R/a$, where R is center-to-center particle separation and A is the Hamaker constant. Steric interactions can be approximated by the hard wall repulsion at 2δ where δ is the thickness of the steric barrier. Finally, magnetic interactions can be described in the first order by dipole-dipole interactions

$$E_M = R^{-3}\left[\vec{\mu}_1 \cdot \vec{\mu}_2 - 3(\vec{\mu}_1 \cdot \vec{r})(\vec{\mu}_2 \cdot \vec{r})R^{-2} \right] \tag{6}$$

where $\vec{\mu}_1$ and $\vec{\mu}_2$ are dipole moments of the particles and $\vec{r}$ is the vector connecting the two dipoles.

In order to gain insight into colloidal interactions, the colloid has to be extensively characterized, determining the parameters in the above expressions. Characterization methods appropriate for nonaqueous colloids are reviewed with emphasis on the most appropriate techniques. These methods are applied to characterization of colloid components, the colloid itself and interactions in the colloid consisting of iron oxide particles, phosphate ester surfactant and binder in cyclohexanone liquid. Extensive characterization of this colloid permits at least elementary understanding of colloidal interactions.

CHARACTERIZATION METHODS

In this section, characterization techniques for powders, solutions, nonaqueous colloids and coatings are reviewed with emphasis on the most relevant techniques. Starting with characterization of powders and their surfaces, particle size and shape can be determined with optical microscopy, secondary electron microscopy and/or transmission electron microscopy. Solid-liquid interactions dominate colloid properties and they are dependent on surface composition and surface chemistry. Ultrahigh vacuum, surface sensitive techniques,[2] such as x-ray photoelectron spectroscopy (xps), Auger spectroscopy and secondary ion mass spectrometry are useful in establishing surface elemental composition and contaminants. In particular, angle resolved xps schematically depicted in Fig. 2 is invaluable in surface characterization of compressed powders and coatings.

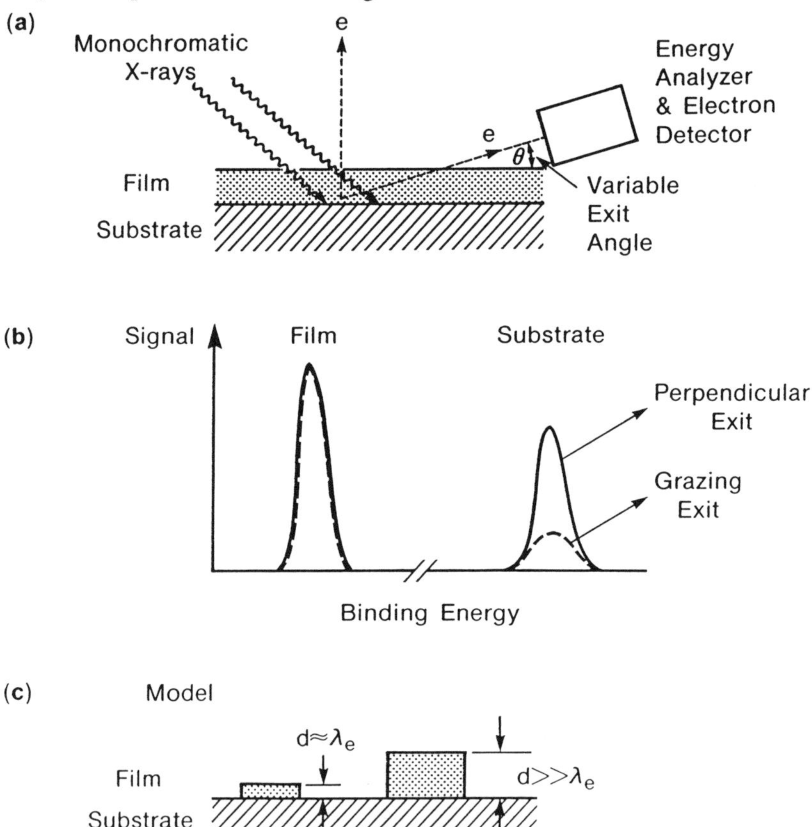

Figure 2: (a) Angle resolved x-ray photoelectron spectroscopy, (b) model of thin film covering the substrate nonuniformly and (c) observed signals from the film and the substrate at grazing and perpendicular exits. Signals from the film were adjusted to the same intensity. λ_e is the escape depth of photoelectrons.

Adsorbed amounts and uniformity of adsorbates can be obtained by simplified analysis.[3] Surface chemistry, specifically basicity and acidity of the surface groups can be determined by surface calorimetry. The technique probes interactions of the surface with any adsorbate by detecting heat evolved or absorbed during physical or chemical interactions. The principle of the technique is shown in Fig. 3a for adsorption of organic molecules with hydroxyl end groups on an iron oxide surface. Heat released or absorbed can be measured with heat flow or adiabatic methods. The schematics of the heat flow calorimetric set up and example of the data and calibration is presented in Fig. 3b. Solid-liquid and liquid-liquid interactions can be monitored with batch, titration or flow through methods.

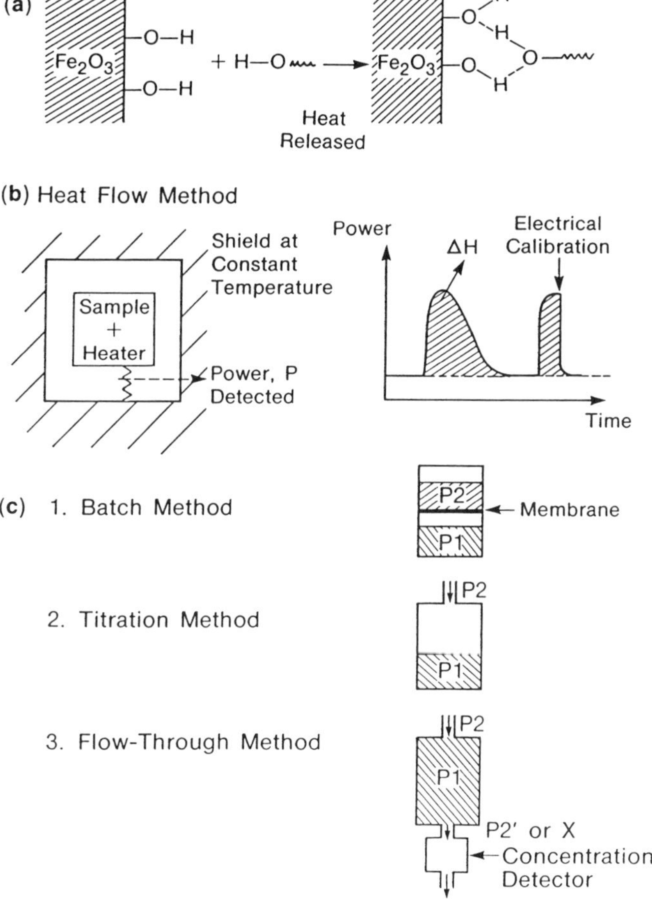

Figure 3: (a) Schematic representation of adsorption with heat released, (b) principle of heat flow calorimetry and example of the data and (c) three methods for study of solid-liquid and liquid-liquid interactions.

The most flexible technique is flow through as heat of dilution, mixing, adsorption and desorption isotherms can be studied in step-wise fashion with variable concentrations and the out flowing liquid can be monitored with a concentration detector, as shown in Fig. 3c. The typical sensitivity of a heat flow calorimeter such as the Bioactivity Monitor[4] is 0.1 μW. With a reaction time of 100 sec, the detectable energy input or output is 10 μJ. The strong solid-liquid interactions generate $\Delta H = 100$ mJ/m^2, and therefore calorimetric techniques have potential for measurements on samples with surface areas down to 10-100 cm^2. When adsorption on solids is studied, corrections for dilutions have to be taken.

Characterization of surfactants, binders, suspending liquids and their solutions should include evaluation of their composition and purity. Technological samples are often composed of many components, are impure and poorly characterized. Nuclear magnetic resonance is useful in determining the molecular structure of these samples and high performance liquid chromatography (hplc)[5] and gel permeation chromatography are appropriate for separation of the different components. Heats of dilution can be measured calorimetrically and surfactant-binder interactions can be examined with the same method. Intensity and quasielastic (dynamic) light scattering can reveal the presence of micelles and their size. Electrical conductivities and electrical transients[6] can yield ionic mobilities and concentrations which are needed for determination of the Debye lengths. Water has often a profound effect on properties of nonaqueous colloids, in particular low conductivity colloids.[6] Concentration of water in solutions, supernatants and dispersions can be measured using K. Fischer titration[7] and supplemented by electrical transients.

Characterization of colloids normally starts with determination of particle size and its distribution. Direct optical measurements for particles above ~1 μm, intensity and quasielastic light scattering, centrifugation methods and sedimentation with optical or x-ray detection are often used. Particle mobility and its distribution can be determined most directly by video microelectrophoresis in high and intermediate conductivity nonaqueous colloids with particle size above ~0.1 μm. Light scattering and electroacoustic methods require deconvolutions to recover information on mobility distribution. An experimental set up with a potential to measure size, mobility and their distributions is shown in Fig. 4. Nonaqueous colloids with low electrical conductivity require special techniques[8] with closely spaced electrodes and optical and electrical transient detection in order to minimize nonuniform and time varying electric fields.

Adsorption isotherms of surfactants, binders and additives are essential in characterization of colloid. They can be determined by measuring concentration of surface active species in supernatant or concentrations of these species directly on particles. Gravimetric, differential refractive index, infrared or ultraviolet absorption, electrical conductivity, etc. are useful, but surface calorimetry, hplc, atomic absorption or plasma emission spectroscopy[9] are often more appropriate. When the concentration of adsorbates is measured on particles, diffuse infrared reflectance, photoacoustic spectroscopy, atomic absorption, plasma emission or surface analysis (e.g., xps) are attractive.

          Characterization Techniques for the Solid-Solution Interface

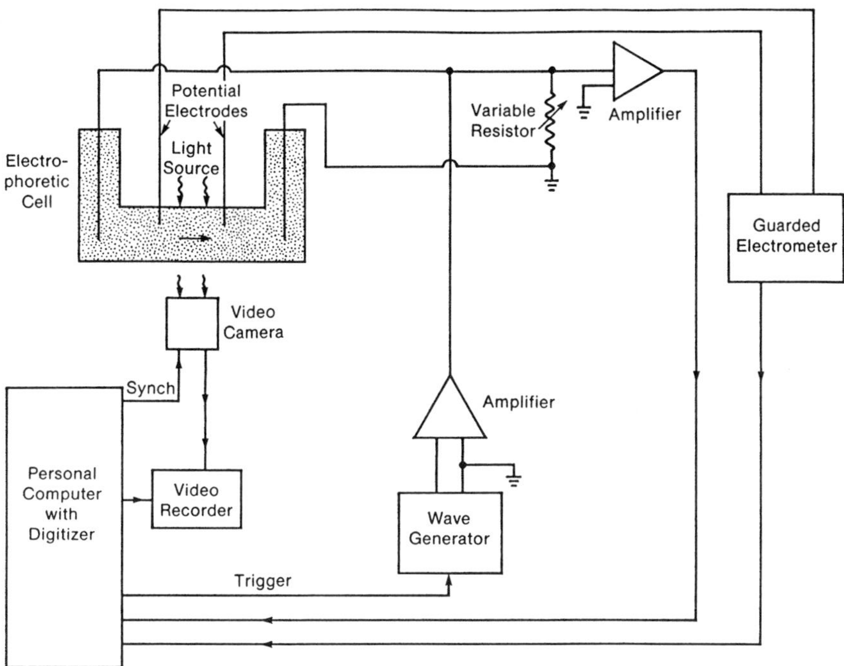

Figure 4: Video electrophoretic set up for colloids with intermediate and high electrical conductivity. Electric field is applied in short (~100's of msec) pulses of reversing polarity. Particle size, mobility, their distributions, electrical transients and electrical field can be measured simultaneously.

Particle-surfactant-binder-liquid interactions can be studied in simple or complex systems by calorimetry. Rheological measurements of yield stress[10] can be performed in concentrated colloids and they sensitively reflect the optimum concentrations of surfactants needed for steric stabilization. The extent of steric interactions can be probed by light and neutron scattering. For magnetic colloids, number of other techniques[11] are useful but those techniques are outside of the scope of this study which emphasizes characterization techniques applicable to all colloids.

Finally, characterization of coatings useful in magnetic recording includes measurements of roughness, thickness, porosity, asperities, defects, pigment volume concentration and mechanical properties - wear rates, hardness, etc. Roughness can be readily measured by mechanical profilometry, optical interference (on samples coated with thin metal layers in order to circumvent the optical nonuniformities) or atomic force microscopy over wide range of spatial frequencies. Porosity can be evaluated by inert gas adsorption or from absorbed lubricant amounts. Asperities are detectable in flyability studies with piezoelectric sensors. In magnetic coatings, defects are mapped by recording tests with flying magnetic head. Pigment volume concentrations can be evaluated by film thickness measurements and gravimetry of pigment after ashing organics. Wear rates are probed with pin-on-disk or slider-on-disk tests and hardness by indentation.

RESULTS AND DISCUSSION

The system on which the above outlined characterization will be illustrated consists of γ-Fe_2O_3, Co doped magnetic particles, small concentration of Al_2O_3 particles, phosphate acid ester surfactant, epoxy-phenolic binder and cyclohexanone, $C_6H_{10}O$, liquid. Iron oxide particles are needle shaped and have average length of 60 nm and average short dimension of 25 nm and their surface area is 30 m^2/gm. Aluminum oxide particles are roughly spherical with diameter of about 200-300 nm. Oxide particles have a considerable amounts of adsorbates on their surfaces as is clear from xps data in Fig. 5a.

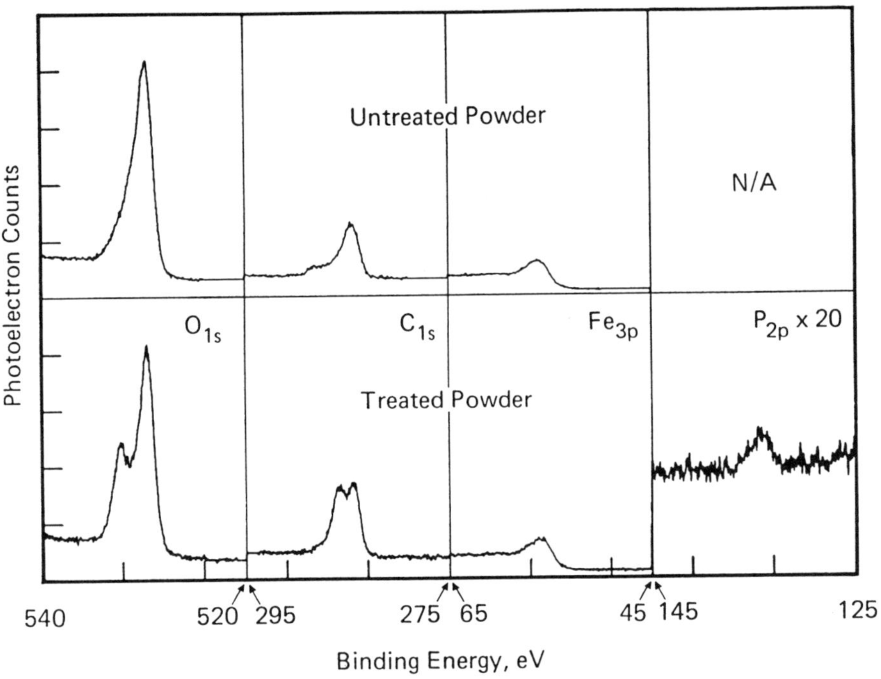

Figure 5: High resolution xps spectra of cobalt doped iron oxide particles before (a) and after (b) surfactant treatment. Each spectrum charges differently and thus spectra were shifted so that experimental C_{1s} peaks from adventitious hydrocarbon on powders coincide with the position of C_{1s} in the reference tables, i.e., 284.6 eV.

    Characterization Techniques for the Solid-Solution Interface

Typical result from surface characterization of iron oxide by calorimetry is shown in Fig. 6. Adsorption and desorption of pyridine which is a basic adsorbate determines the surface density and strength of acidic groups.[12,13] For probing of basic groups on the surface, acidic liquid (n-butyl alcohol) is employed. Data is summarized in Table 1 for cleaned, as received iron oxide and aluminum oxide. It is clear that as received powder is very different from powder which was cleaned by Soxhlet extractions with several solvents. Iron oxide has similar acidity and basicity and concentration of surface groups as measured by heat of adsorption.

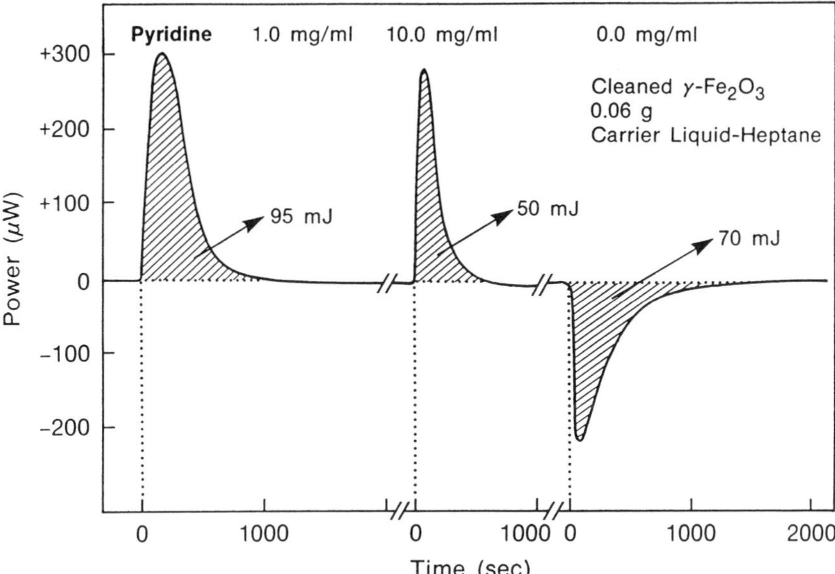

Figure 6: Heat of adsorption of basic liquid (pyridine) from inert liquid (heptane) onto iron oxide surface at two concentrations of pyridine, followed by desorption.

Table 1. Heat of adsorption and desorption, ΔH.

Sample	Acidic ΔH (mJ/m^2)		Basic ΔH (mJ/m^2)	
	Adsorption	Desorption	Adsorption	Desorption
Cleaned γ-Fe$_2$O$_3$	68 $\pm$12	45 $\pm$9	60	60
As received γ-Fe$_2$O$_3$	120	80	130	130
Aℓ_2O$_3$	85	80	150	120

Surfactant-phosphate acid ester (PE) of ethoxylated alcohol, RE-610 was found to be a mixture[14] of monoester and diester by proton nuclear magnetic resonance. Monoester and diester have the following formulae:

$$
\begin{array}{ccc}
\text{O} & & \text{O} \\
\| & & \| \\
\text{R}-\text{P}-\text{OH} & \text{and} & \text{R}-\text{P}-\text{R} \\
| & & | \\
\text{OH} & & \text{OH}
\end{array}
$$

where $R = CH_3(CH_2)_8 - \langle \overline{O} \rangle - O - (CH_2CH_2O)_n$. Molecular weight and concentrations of monoester and diester are 696 and 62 wt% and 1294 and 38 wt% with $n \approx 9$, respectively. The stretched monoester molecule has an estimated length of 5 nm. The electrical transients of PE in cyclohexanone yield an electrical conductivity of 1.6×10^{-3} S/m at concentrations of 50 mg/ml and ionic mobilities around $0.22 \times 10^{-8} m^2/V$ sec.

Interaction of PE surfactant with iron oxide particles is reflected in the flow calorimetric data shown in Fig. 7. The heat of adsorption is $\sim$10 mJ/m^2 and is almost completely reversible indicating that the surfactant is only weakly physisorbed on the oxide surface. Similar experiments with epoxy - phenolic show even weaker interaction with the oxide surfaces, $\sim$1 mJ/m^2. Epoxy does not appear to interfere with the previously adsorbed surfactant.

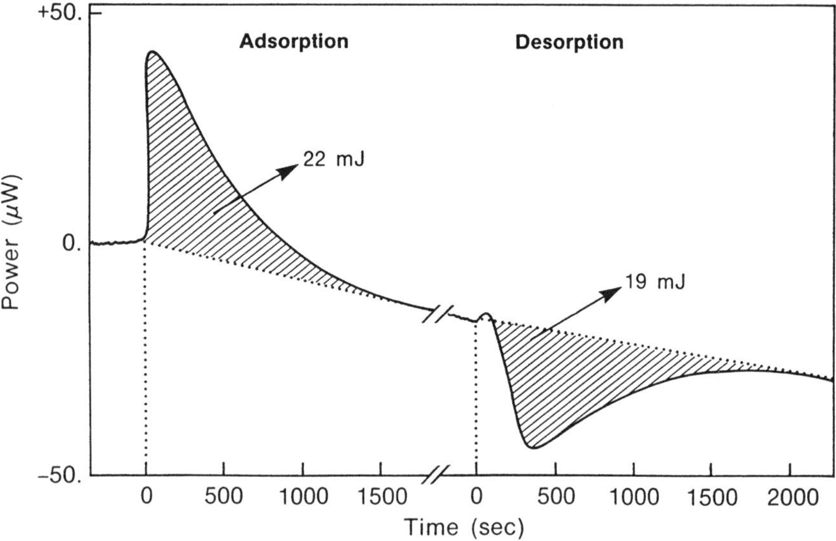

Figure 7: Adsorption and desorption of PE surfactant on iron oxide surface in cyclohexanone measured calorimetrically.

    Characterization Techniques for the Solid-Solution Interface

The optimum concentration of surfactant needed for the best dispersion can be determined from the adsorption isotherm. The data in Fig. 8 is obtained from plasma emission spectroscopy of phosphorous remaining in the supernatant after adsorption on the iron oxide particles. The adsorbed amount saturates at about 4 mg of phosphorus per 1 gm of iron oxide. If this amount corresponds to 1 monolayer coverage, then PE molecule occupies about 0.8 nm^2 and has likely normal orientation to particle surfaces.

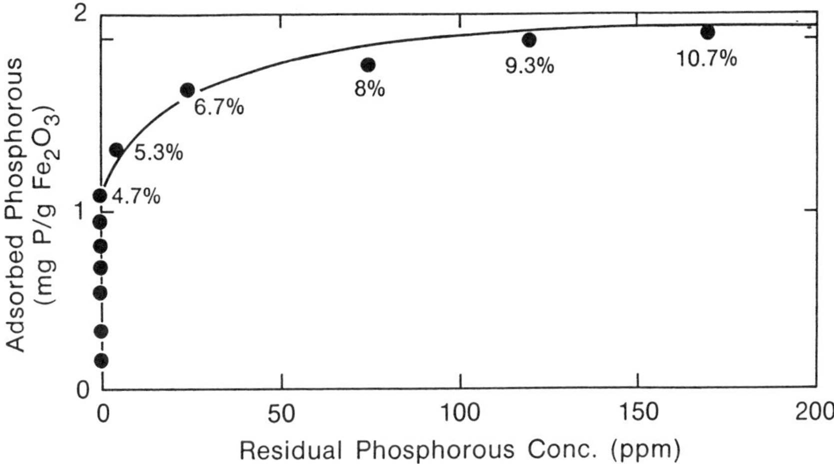

Figure 8: Adsorption isotherm of PE surfactant on iron oxide measured by plasma emission spectroscopy of phosphorus in the supernatant.

Dispersions are prepared by ball milling of powder with PE in cyclohexanone using 25 wt% of iron oxide. When needed, dilutions are made into the supernatant and dispersions are sonified before measurements. With optimized surfactant concentration, the average particle size is 0.2 μm, much greater than the primary particle size even shortly after redispersion. The agglomerates have wide particle distribution and their size increases with time after redispersion. In Fig. 9, the electrical conductivity of PE solution and the electrophoretic mobility of particles are shown as a function of the surfactant concentration. The electrical conductivity rises linearly with increasing PE concentration above cyclohexanone conductivity. Iron oxide particles in cyclohexanone are negatively charged, while small concentrations of PE reverse charge polarity. The positive charge on particles saturates at about 1.5×10^{-9} m^2/V sec which is lower in absolute value than mobility in solvent without surfactant. If positive charge originates from ionization of hydroxyl groups of surfactant, the excess of hydrogen ions exists on iron oxide surfaces.

Concentrated dispersions can be conveniently examined in rheological measurements[14] of yield stress. Yield stress goes through a minimum at about 6.5 wt% of PE with respect to iron oxide weight which also defines the concentration of surfactant for optimum dispersion. This concentration is in a good agreement with adsorption isotherm in Fig. 8.

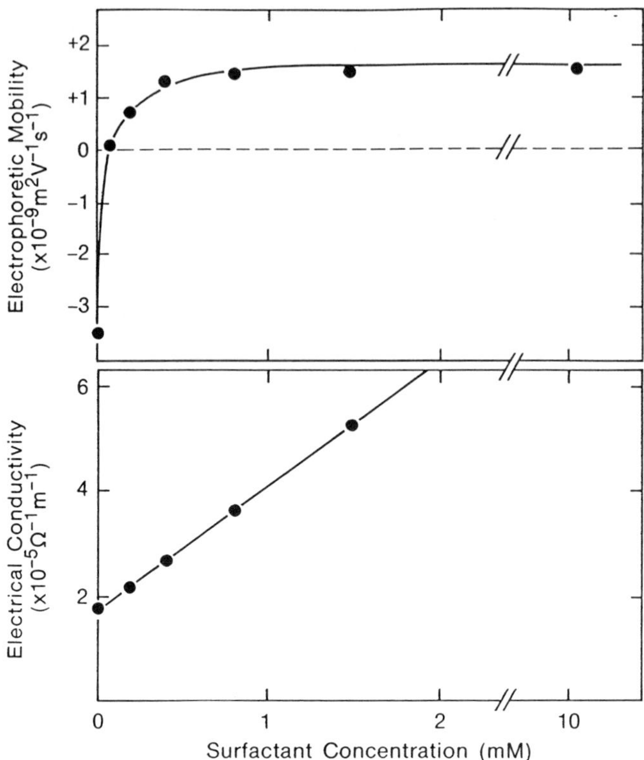

Figure 9: (a) Electrophoretic mobility of iron oxide particles in PE + cyclohexanone solution and (b) electrical conductivity of this solution.

Particulate coatings produced with these colloids have a rms roughness of 100 nm after curing and buffing. They have high porosity, 20-30% by volume as determined from lubricant absorption.[15] Porosity is deliberately high so that a large amount of lubricant can be accommodated in the coating and wear resistance can be significantly improved, about 1000 times compared with an unlubricated coating. The pigment volume concentration is high, around 28 ± 3 volume percent. After buffing, flying can be performed reliably down to 100 nm, indicating that asperity heights are lower than this value.

It is interesting to study what happens to the surfactant after curing the coating at 220°C. Xps data speaking to this issue are given in Fig. 10 and Table II. Curing lowers P/Fe somewhat, but a significant amount of surfactant remains in the coating. Also humidity exposure does not change the P/Fe ratio. Presence of Si on the uncured coating is due to a surface active additive in the colloid. This material is confined only to the surface and is removed almost completely with buffing.

     Characterization Techniques for the Solid-Solution Interface

 Elemental composition of treated and untreated iron oxide powders and magnetic particulate coatings (at%)

	O	C	Al	Si	P	Fe	P/Fe
Untreated oxide	51.3	38.9	--	--	--	9.8	--
Treated oxide	43.5	47.5	--	--	0.8	8.1	0.10
Uncured coating	25.9	66.7	--	5.4	0.18	1.8	0.10
Cured coating	29.5	65.0	--	1.2	0.29	3.9	0.07
Buffed coating	28.0	67.8	0.3	0.13	0.19	3.5	0.05
Post-humidity coating	26.5	69.7	--	--	0.20	3.6	0.06

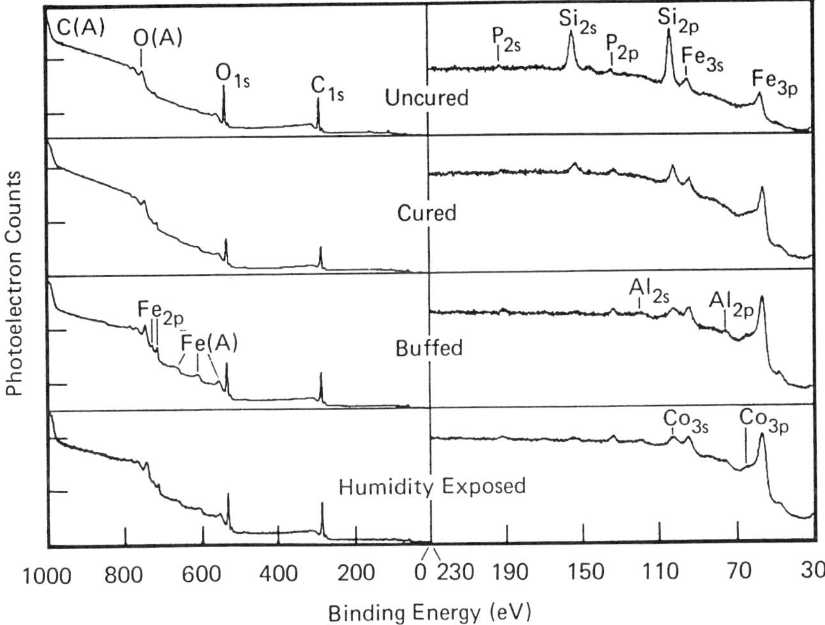

Figure 10: Survey x-ray photoelectron spectra and 200 eV spectra of coatings in the conditions indicated.

Angle resolved xps can be used to assess how iron oxide is embedded in the binder. Data in Table III show that only a small fraction of iron oxide is not covered or thinly covered with the binder. The apparent bare fraction can be partially or totally due to surface roughness. The estimated electron mean free path (mfp) in the organic overcoat is of the order of 3 nm.

Characterization Techniques for the Solid-Solution Interface

Mechanical wear resistance of this coating is adequate, as it takes typically 500 sliding cycles to generate an optically observable wear track,[16] as shown in Fig. 11.

Table III. Surface analysis of particulate coating

Percentage of surface which is:			
Bare Fe_2O_3	Thinly Covered Fe_2O_3	Thickly Covered Fe_2O_3	Thickness of Thin Layers over Fe_2O_3 (Electron mfp's)
3	5	92	0.7

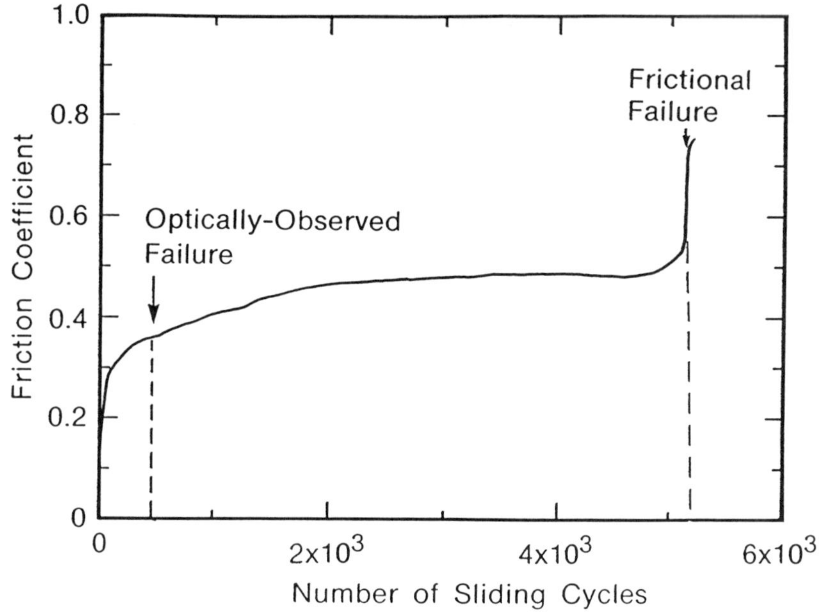

Figure 11: Tribological data from a sliding experiment with the 10 cm dia spherical cap pin loaded to 0.15N at 1m/sec sliding velocity. Optically detectable wear corresponds to the initial wear while the frictional failure marks the number of sliding cycles needed to wear the coating and expose the substrate.

With the above data, colloidal interactions can be summarized for individual primary and agglomerated particles. Particle mobilities were measured only on agglomerated particles which have about 25 elementary charges and corresponding ζ potentials of 25 mV. It is assumed here that primary particles have the same surface potential as agglomerated particles. The concentration of ionizable species in solution at 1 monolayer PE coverage of iron oxide is 2×10^{20} m^{-3} from electrical transient analysis. For PE, valencies z_i's in Eq. (1) are likely 1, and therefore $\lambda = 15$ nm, which is indeed

    Characterization Techniques for the Solid-Solution Interface

smaller than dimensions of primary particles and much smaller than average radius of agglomerated particles. Electrostatic interactions from Eq. (4) for primary colloidal particles are shown in Fig. 12. Van der Waals interactions given by Eq. (5) with $A = 5 \times 10^{-20}$ J are also included in Fig. 12 together with hard steric wall of $\delta = 5$ nm represented by the length of the surfactant molecule.

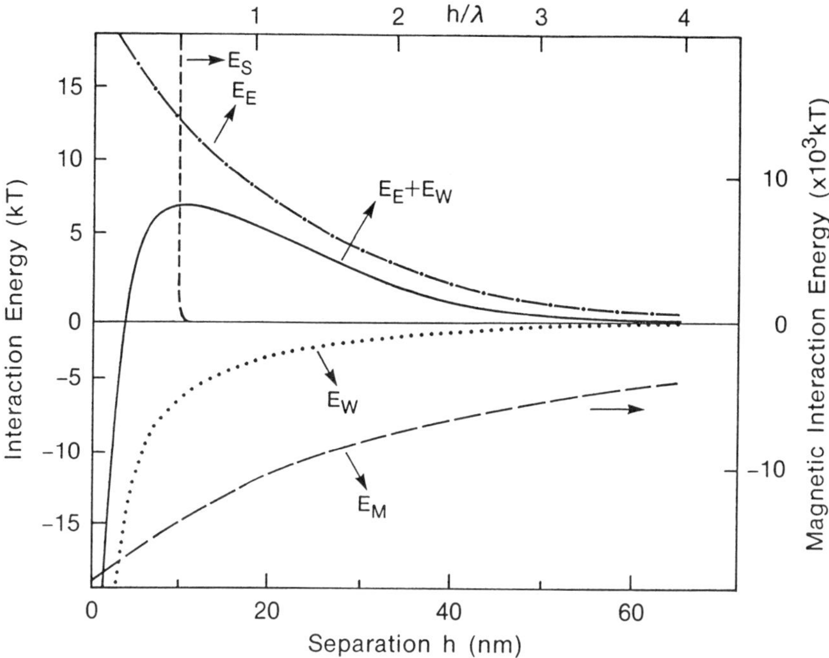

Figure 12: Electrostatic, E_E, van der Waals, E_W, steric, E_S and magnetic E_M interaction energies vs particle separation, h, and vs h/λ for studied colloid.

For antiparallel dipoles the magnetic interaction from Eq. (6) simplifies to $E_M = -\mu^2/R^3$. Magnetic dipoles of particles are proportional to their saturation magnetization, M_s, and particle volume, V, consequently $E_M \sim a_{eff}^6$, the sixth power of the effective particle size of the individual particles ($a_{eff} \approx$ 20nm). The magnetic attraction of primary particles in Fig. 12 completely overwhelms the repulsion for all particle separations except below the hard steric wall. In agglomerated particles, overall magnetic moment is much smaller than the sum of moments of individual particles and magnetic attraction appears to be of the same order as repulsion as slow flocculation of agglomerates indicates.

SUMMARY

Magnetic attractions predominate in these iron oxide based colloids and consequently these colloids are unstable. Smaller particles are required in order to decrease magnetic dipole-dipole interactions. Electrostatic stabilization is weak in these systems and electrostatic repulsion is weakened by the addition of phosphate ester surfactant which reverses the original particle charge. Interaction energy of phosphate ester surfactant with iron oxide in a relatively polar liquid is low and adsorption is reversible to a large extent. Electrostatically, the colloid can be characterized as an intermediate conductivity colloid with an electrical double layer shorter than the particle size. Better steric stabilization is essential for improvement of stability of these colloids. Improved steric stabilization would require chemical grafting or block copolymer adsorption. Adequate thin particulate coatings are produced, but more stable colloids would yield improved particulate coatings.

REFERENCES

1. P. H. Wiersema, A. L. Loeb and J. Th. G. Overbeek, *J. Colloid Interface Sci.* **22**, 78 (1966).

2. D. Briggs and M. P. Seah, "Practical Surface Analysis by Auger and X-ray Photoemission Spectroscopy," J. Wiley, New York, NY (1983).

3. C. M. Mate and V. J. Novotny, *J. Chem. Phys.* **94**, 8420 (1991).

4. J. Suurkuusk and I. Wadso, *Chemica Scripta* **20**,155 (1982).

5. S. G. Perry, R. Amos and P. I. Brewer, "Practical Liquid Chromatography," Plenum (1972).

6. V. J. Novotny, *J. Electrochem. Soc.* **133**, 1629 (1986).

7. I. Kolthoff, "Treatise on Analytical Chemistry," J. Wiley, Part II, Volume 1, Section A, 69 − 165 (1961).

8. V. J. Novotny, *J. Appl. Phys.* **50**, 324 and 2787 (1979).

9. P. W. Boumans, "Inductively Coupled Plasma Emission Spectroscopy," Wiley (1987).

10. J. C. Patton, "Paint Flow and Pigment Dispersion," J. Wiley, Interscience, 297 (1979).

11. S. Dasgupta, *IEEE Trans. Mag.* **MAG-20**, 7 (1984).

12. R. S. Drago, L. B. Parr and C. S. Chamberlain, *J. Amer. Chem. Soc.* **99**, 3203 (1977).

13. F. M. Fowkes, H. Jinnai, M. A. Mostafa, F. W. Anderson and R. J. Moore "ACS Symposium Series," **200**, 307 (1982).

14. R. Hannan, private communications.

15. M. R. Lorenz, V. J. Novotny and V. R. Deline, *IEEE Trans. Mag.* **27**, 5088 (1991).

16. T. E. Karis, V. J. Novotny and R. M. Crone, *ASLE (SP-29)*, 35 (1990).

Particle Attraction Effects on the Centrifugal Casting and Extrusion of Alumina

C. H. Schilling
Ames Laboratory* and Department of Materials Science and Engineering
Iowa State University
Ames, Iowa 50011-2070

L. Bergström
Institute for Surface Chemistry
Stockholm, Sweden

H.-L. Ker and I. A. Aksay
Department of Chemical Engineering and Princeton Materials Institute
Princeton University
Princeton, New Jersey 08540-5211

ABSTRACT

Interparticle attraction forces were empirically related to the centrifugal casting and extrusion behavior of flocculated alumina suspensions. Attractive forces were altered by two approaches: (*i*) salt flocculation, which entails regulating the electrical double-layer thickness through electrolyte additions; and (*ii*) screening of van der Waals attraction by the steric interactions of surface-adsorbed fatty acids. Specimens produced by both methods exhibited a shift from high to low

---

* Ames Laboratory is operated for the U.S. Department of Energy by Iowa State University under contract No. W-7405-Eng-82.

compressibility at a critical maximum density that increased with decreasing interparticle attraction. Gradients in packing density during centrifugal casting were alleviated using methods (*i*) and (*ii*) as long as spatial variations of the effective stress were within the low-compressibility range. We hypothesized that the reduced interparticle attraction in methods (*i*) and (*ii*) may also raise the threshold packing density at which ductile-to-brittle transitions occur during plastic shear. Specimens prepared with oleic acid adlayers were highly plastic and easily extrudable at solids contents of up to 59 vol%, although salt-flocculated samples at 55 vol% density extruded at creeping flow rates that were insensitive to the applied pressure. Results suggested that particle rearrangement during shear is a rate-limiting process, with an average relaxation time that is lowered by reducing interparticle attraction.

INTRODUCTION

It is important to regulate spatial variations of packing density that arise in the slurry consolidation of ceramic powders since they can lead to shape distortion and cracking during drying, pyrolysis, and sintering, or they can produce fracture origins in sintered materials. These packing density gradients are caused primarily by nonuniform effective stresses (i.e., local compaction stresses) that are inherent in a particular liquid removal process.[1-5] For example, in filtration, drag forces originate at the pore walls and are transmitted through interparticle contacts such that the effective stress increases in the direction of flow.[1-3] In centrifugal casting, packing density gradients result from local variations in both the radial acceleration and the buoyed weight of a particle network.[3-5]

Gradients in effective stresses are difficult to eliminate in centrifugal casting, because of fluid mechanics considerations which are discussed in this paper. Despite these stress gradients, it is possible to avoid packing density gradients by manipulating colloidal interactions so as to reduce the compressibility of a particle network. This is traditionally accomplished by developing strongly-repulsive interparticle forces and avoiding strongly-flocculated, compressible

 Characterization Techniques for the Solid-Solution Interface

slurries.[6] Another alternative is to use weakly-flocculated suspensions that exhibit a transition from high to low compressibility above a critical value of the effective stress.[2,3,7] Within the slurry, the effective stress must everywhere be above this critical value in order to achieve uniform densities in a given shaping operation. In this study, we analyze the effects of varying the degree of flocculation on packing density gradients during centrifugal casting. We address this problem by comparing the compressive yield behavior of model alumina suspensions for which interparticle attraction forces are systematically varied through changes in the suspension chemistry.

Regulating interparticle adhesion forces using model suspensions is also helpful in developing a basic understanding of plasticity control in shear-forming applications. For example, a significant problem in shear-forming nonclay ceramics is that relatively large concentrations of interparticle liquids are typically required to maintain tractability and avoid a transition to brittle, dilatant behavior that occurs above a critical solids concentration (40 to 55 vol% for monosized, spherical particles).[8] However, shape distortion and cracking are frequently attributed to internal stresses that are caused by the removal of such large amounts of liquids during drying.[9] In addition, drying cracks are generally more severe as the size and complexity of a shaped article increases. Although this problem is common to both clays and nonclays, the latter are frequently more difficult to process into crack-free shapes exceeding 20 to 30 centimeters in size.

Exact reasons for the differences in the plasticity of clays and nonclays are not well understood, although a literature review suggests that they are likely attributable to variations in inherent micromechanisms for elastic strain energy relief.[3] It may be argued that clays are more plastic than nonclays because of their stacked platelet morphology and that platelet sliding is promoted by a combination of long-range attraction (van der Waals) and short-range hydration repulsion associated with surface-adsorbed water molecules.[10] Nonclay powders are usually more equiaxed in shape, and a similar combination of attractive and repulsive surface forces that aid particle rearrangement is only made possible in special cases by using appropriate surfactants.

Additional research is needed to explore how surfactant chemistry and interparticle attraction may be regulated to achieve high density, more tractable nonclay suspensions for shear-forming applications. We address this problem using simple extrusion experiments with the model suspensions described below. In addition, we explore how variations in the suspension chemistry influence shape retention and creep relaxation behavior.

EXPERIMENTAL PROCEDURE

Samples were flocculated in two different ways that allow systematic variations in interparticle attraction to be made: salt flocculation[11] and the surface adsorption of fatty acids.[3,4,12] As a comparison, suspensions that were strongly-flocculated near the isoelectric point were also prepared. A submicron powder produced by spray pyrolysis of alkoxide droplets was chosen to prepare all suspensions (Sumitomo Chemicals, Osaka, Japan, Type AKP-30). The particle size distribution of a stable aqueous dispersion, measured by a sedimentation technique that uses x-ray absorption (Sedigraph 5000 E, Micromeritics, Norcross, GA) showed a mean particle diameter of 0.4 μm with 90% of the distribution within the size range of 0.25 to 1.2 μm.

Salt-flocculated suspensions were prepared by adding as-received powder to distilled water and electrostatically stabilizing at pH 4.0 using HCl additions. These pH 4 suspensions were ultrasonicated to break up agglomerates. Aqueous salt solution was subsequently added to raise the concentration in the suspending medium above the critical coagulation concentration (≈ 0.1 M). Two separate sets of samples were prepared by adding solutions containing either NaCl or NH_4Cl. In the latter case, small amounts of HCl were simultaneously added with NH_4Cl to maintain the pH at 4.0. The addition of salt is believed to establish a short range repulsion force that promotes the rearrangement of particles.[11] The origin of this repulsion is unclear, but it has been suggested that it is caused by short range double-layer effects or by the structuring of water molecules between two particles in close contact.[13] Precise estimates of the net interparticle potential are

                    Characterization Techniques for the Solid-Solution Interface

not available for these systems, although double-layer theory suggests that raising the electrolyte concentration decreases the double-layer thickness, which in turn strengthens interparticle attraction.[14]

Aqueous suspensions that were strongly-flocculated near the isolectric point were also prepared using the above procedures up to and including HCl addition and ultrasonication at pH 4.0. NH_4OH was subsequently added to raise the pH to 8.6.

Suspensions containing fatty acids were prepared by mixing as-received powder with a solution of monodispersed fatty acids and decalin solvent. We chose this model system to simplify calculations of the interparticle potential by avoiding electrostatic interactions and thereby reducing the components of the net surface force to van der Waals attraction and short-range steric repulsion.[4,12] The fatty acids are linear alkanes having a carboxylic acid head group that hydrogen bonds to Al-OH groups on the alumina and promotes the extension of each molecule roughly perpendicular to the surface (Figure 1). Based on adsorption isotherm analysis, a sufficient amount of fatty acid was added to each suspension to saturate the particle surfaces.[4,12] We also prepared a series of suspensions containing fatty acids of increasing molecular weight, and the adlayer thickness was estimated as approximately equal to the length of the acid molecule in each case. Theoretical estimates of the interparticle potential revealed that increases in the adlayer thickness screen the van der Waals attraction and raise the minimum adhesive energy between particles (Figure 1).[4,12]

Flat-bottomed, poly-acrylate tubes measuring 5.72 and 7.62 cm inside diameter respectively were used for gravity settling and centrifugation experiments. Density mapping by gamma-ray densitometry[3,15] was performed to evaluate consolidation behavior. For these experiments, a tube was placed in an upright position and irradiated by a horizontal, 3.2-mm-diameter photon beam passing through the central axis of the tube (661 keV from cesium-137). The Beer-Lambert law was applied to photon attenuation measurements at several elevations in the tube and used to obtain density profiles (accuracy $\pm$ 1.5 vol%). Measurements were performed in real time during sedimentation until the density profiles were time invariant. For the centrifugation experiments, density mapping

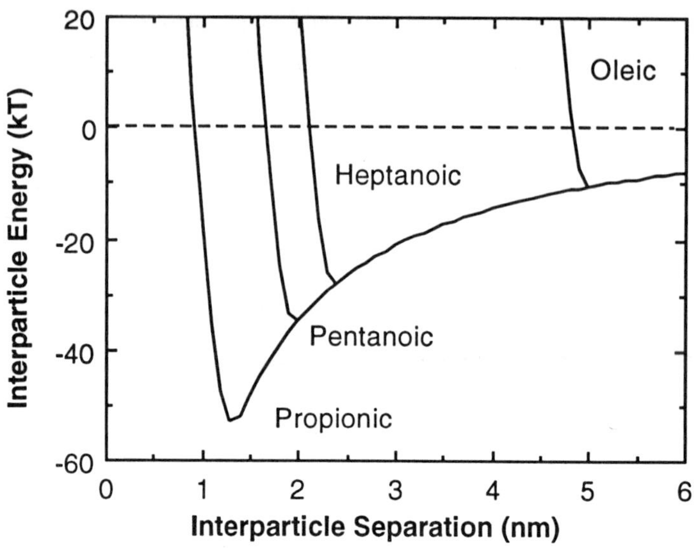

Figure 1. At the top, calculated interparticle energies are shown for 0.4 μm diameter alumina particles in decalin and coated with fatty acids of varying molecular weight.[4] Below are the chemical structures of each fatty acid.

was performed immediately after removing samples from the centrifuge. We repeated this procedure at various rotational speeds and time durations to achieve maximum densification.

Yield properties of sedimentation cakes in drained compression were analyzed by densitometry and fluid mechanics modeling.[3-5,7] The model assumes that, during consolidation, networks of flocculated particles form an elastic continuum that transmits the weight, buoyancy, and viscous drag forces from one elevation to the next. The net force transmitted by the network at a particular elevation is equal to the dynamic balance of the above forces summed over the entire network above that elevation. Shear stresses acting between the cake and the container walls are neglected, and point contacts between all particles are assumed. For cakes that have reached maximum densification (i.e., stable cakes), the viscous drag component is zero, and the effective stress profile is calculated by the following formula:

$$\sigma'(z^*) = \int_{z^*}^{z_{max}} \Delta\rho \, G \, \phi \, dz \tag{1}$$

where $\sigma'(z^*)$ is the effective stress at a perpendicular distance of z^* from the bottom of a cake (i.e., the bottom of a poly-acrylate tube), z_{max} is the total cake height, $\Delta\rho$ is the mass density difference between the solid and liquid, and ϕ is the local volume fraction of solids. G is the local acceleration, which is 981 cm/sec^2 for gravity settling and ω^2 (R_{max} - z) for centrifugation; ω is the rotor angular velocity and R_{max} is the distance from the rotor center to the bottom of each cake (20.4 cm). Equation (1) is solved by numerical integration of the density profiles for each stable cake.[3,4]

A piston extruder was fabricated from stainless steel and used to evaluate the shear behavior of various samples (Figure 2). Extrusion was performed under constant stress conditions, and the steady-state velocity of the ram was electronically recorded as a function of the applied load. Slurry preparation closely followed the above procedures, and batch samples having controlled uniform density were generated by constant pressure filtration.[3]

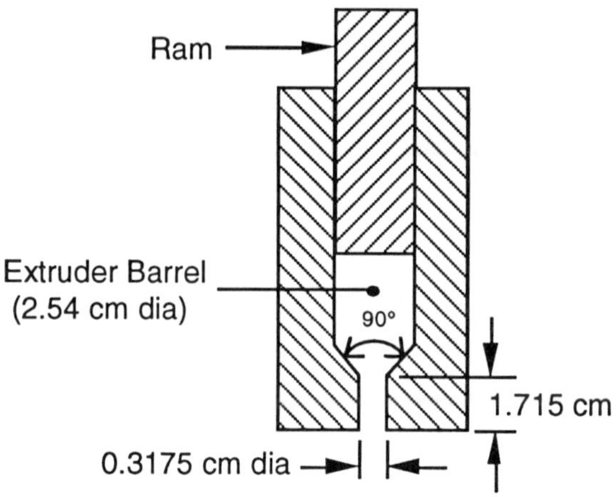

Figure 2. Schematic Illustration of the Extruder. A constant load is applied to the ram, and the ram displacement is continuously measured.

YIELDING IN DRAINED COMPRESSION

All of the stable cakes produced at pH 8.6 (strongly-flocculated slurries) exhibited a gradient towards larger densities and effective stresses when passing from the top to the bottom of a sample (Figure 3). The gravity sedimentation cake exhibited the lowest packing density, whereas increasing the centrifugal acceleration produced successively greater effective stresses and densities at a given elevation. It should also be noted from Figure 3 that calculated local values of effective stresses increased by as much as an order of magnitude from the top to the bottom of a sample.

Stress and density values at each elevation in Figure 3 are combined in Figure 4, which provides a convenient representation of the effective yield stress needed to achieve a given packing density for this slurry system. It is apparent that the strongly-flocculated pH 8.6 samples exhibited a gradual decrease in compressibility with increasing effective stress, and there is good agreement with the consolidation data reported earlier by Lange and Miller[6,16] for a similar α-Al_2O_3 powder prepared at pH 8.6 (Figure 4).

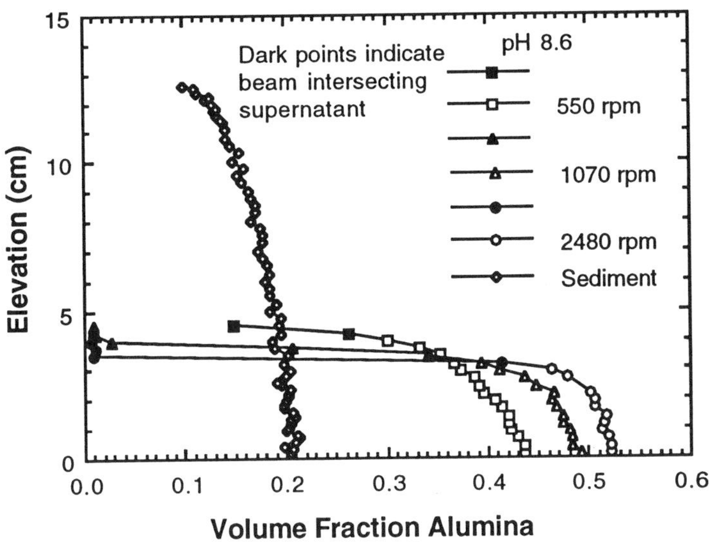

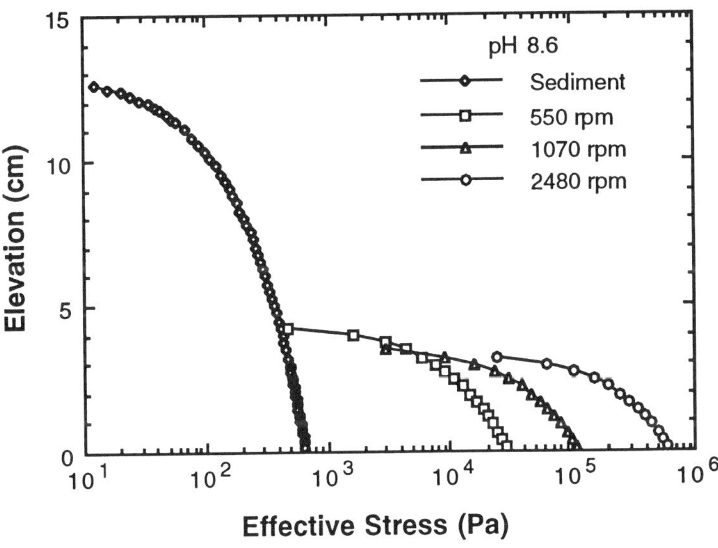

Figure 3. Density Profiles (top) and Corresponding Effective Stress Profiles (bottom) for Stable Cakes Prepared by Sedimentation and Centrifugation of Strongly-Flocculated Alumina at pH 8.6.

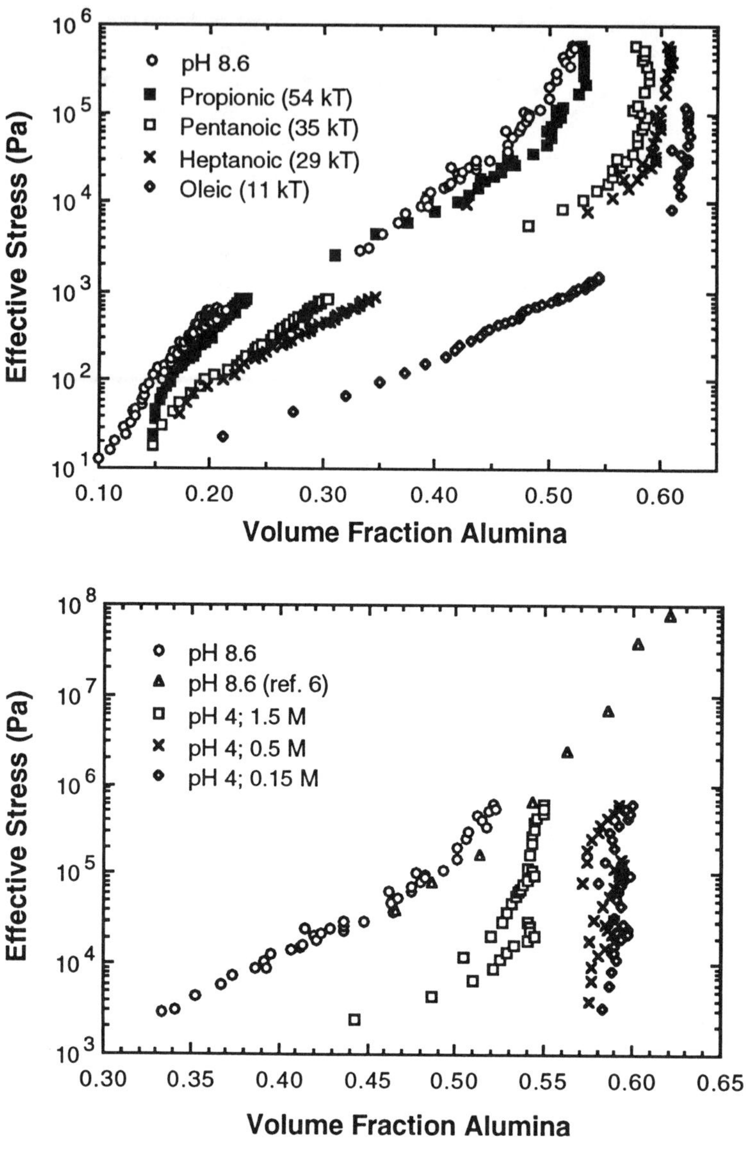

Figure 4. Summary of Stress-Density Data. The samples containing fatty-acids (top) and the salt-flocculated samples prepared with ammonium chloride (below) exhibit a transition from high to low compressibility at a maximum packing density that increases with decreased interparticle adhesion.

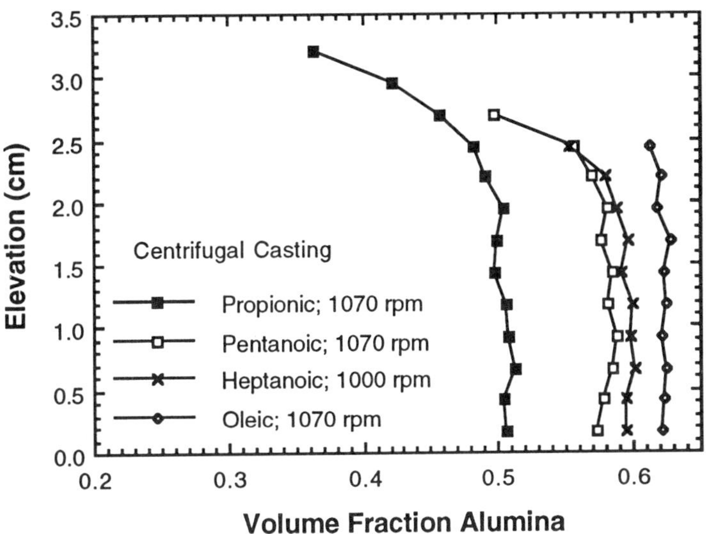

Figure 5. Equilibrium Density Profiles for Stable Alumina Cakes Prepared by Centrifugation. Higher average densities and more uniform densities are obtained by decreasing interparticle attraction. Data points are omitted where the photon beam intersected the cake-supernate interface.

Also shown in Figure 4 are the results for the samples prepared with fatty acids and by salt flocculation with NH_4Cl. In contrast to the strongly-flocculated pH 8.6 samples, the remaining samples exhibited transitions from high to low compressibility as the effective stress increased. A striking feature of both the salt-flocculated systems and the fatty-acid systems was that the densities of the low-compressibility states were increased by reducing interparticle attraction. Solids densities of up to 60 vol% were obtained using the most weakly-flocculated systems containing the longest fatty acid or the least amount of salt. In addition, salt-flocculated samples prepared with either NaCl or NH_4Cl as the additive exhibited nearly identical trends in stress-density behavior.[3]

Density profiles from the centrifugal casting experiments strongly indicated the importance of minimizing interparticle attraction to promote low compressibility and uniform consolidation. As shown in Figure 5, greater average densities and more uniform packing resulted from increasing the molecular length of the adsorbed fatty acid. Essentially the same behavior was noted for samples prepared by both methods of salt flocculation, i.e., the lesser the salt content, the more uniform the density profile in centrifugal casting.[3]

An important practical consideration of the above densitometry-fluid mechanics analysis is that it provides compressibility data below 10 kPa, which is approximately at the lower limit of pressure in traditional consolidation tests.[17-19] Based on the results in Figures 3 and 4, it is apparent that drained compression data are needed at the much lower stresses which exist in gravity sediments and centrifuge cakes.

EXTRUSION

Although we illustrated surfactant systems that produce highly dense and uniform packing in consolidation, there are two additional requirements for plastic shaping applications such as extrusion: (*i*) rapid, plastic flow at low stresses during the shaping process, and (*ii*) shape retention in subsequent handling operations. We performed additional experiments to rank the ability of the above slurry systems to meet these criteria.

As shown in Figure 6, the oleic acid systems meet the first requirement, as they exhibited a steep increase in extrusion rate with increasing applied stress, despite their high packing densities of up to 59 vol%. In contrast, salt-flocculated samples at 55 vol% density exhibited creeping flow extrusion (ram speeds below 0.5 cm/min), with no apparent dependence of the extrusion rate on the applied stress. Strongly-flocculated samples at pH 8.6 were barely extrudable at a density of 55 vol% and exhibited shear-hardening (diminishing ram velocity at each applied stress). Lowering the density of the pH 8.6 systems to 50 vol% produced samples that readily extruded.

Although data in Figure 6 suggests that the oleic acid systems may have poor shape retention, the opposite behavior was observed in qualitative, hand-molding tests that were performed on the extrusion batch samples. The oleic acid samples exhibited a definite yield strength when loaded in tension and a plastic,

          Characterization Techniques for the Solid-Solution Interface

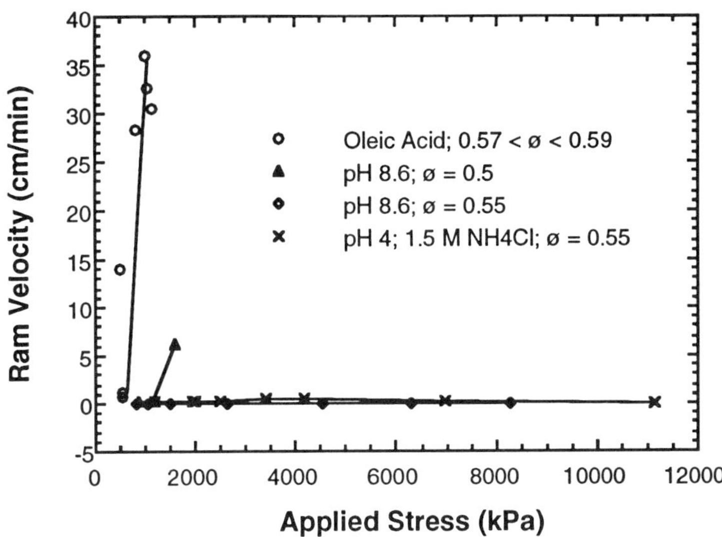

Figure 6. Extrusion Data Summary. Oleic-acid systems exhibit rapid extrusion, whereas the salt-flocculated samples extrude at creeping flow rates. Strongly-flocculated alumina at pH 8.6 loses its extrudability when the packing density is raised from 50 to 0.55 vol%.

clay-like consistency at higher stresses. Similar behavior was exhibited by the strongly-flocculated pH 8.6 sample at 50 vol% density. Raising the density of the pH 8.6 sample to 55 vol% significantly increased the yield strength and produced a transition from a mass with a plastic, clay-like consistency to one exhibiting brittle fracture upon tensile straining. At low tensile strain rates the salt flocculated samples at 55 vol% density exhibited minimal yield strengths and highly-plastic, creeping flow behavior. Raising the applied tensile stress produced a stiffer consistency in the salt-flocculated samples, with the strain rate being relatively unaffected by the stress, and brittle fracture occurring at the highest stresses.

After the density mapping experiments the wet centrifuge cakes were removed from their poly-acrylic tubes and placed intact on a laboratory bench for visual observation. The fatty-acid systems and the strongly-flocculated system at pH 8.6 exhibited minimal shape distortion. Only the salt-flocculated samples exhibited creep deformation under the action of their own self-weight. The rate and extent of creep deformation increased in samples containing lower concentrations of salt, suggesting that particles in these cases were freer to rearrange themselves despite their higher average packing densities. This trend also appeared in specimens prepared with both NaCl and NH_4Cl as the additive. These results clearly show that the elastic strain energy associated with supporting the weight of each sample is more readily dissipated with decreasing interparticle attraction.

We do not fully understand the microscopic relaxation mechanisms causing creep at low stresses in the salt-flocculated systems and insignificant creep at approximately the same stresses in the remaining samples. Also not completely understood is why the extrusion rates increase at a given applied stress when passing from (*i*) the strongly-flocculated pH 8.6 samples at 55 vol% density, to (*ii*) the salt-flocculated samples at the same 55 vol% density, to (*iii*) the oleic acid samples at higher densities. Plastic flow appears to be controlled by rate-limiting micromechanisms having an average relaxation time that decreases when passing from system (*i*) to (*ii*) to (*iii*). This general principle is used to explain ductile-brittle transitions in polymers and crystalline solids, although the rate-limiting micromechanisms differ.[20-21]

Particle rearrangement and the viscous flow of interparticle fluid are primary micromechanisms of strain energy relaxation in dense colloidal systems. Plastic flow rates for a given loading condition are largely dependent on particle mobility, which is in turn affected by the packing density and the magnitude of interparticle friction and adhesion.[22] Although calculated interparticle forces are not available for all of the systems tested, we believe that the average interparticle attraction is reduced when passing from system (*i*) to (*ii*) to (*iii*). In turn, this reduced interparticle attraction may explain the systematic increases in the extrusion rates.

        Characterization Techniques for the Solid-Solution Interface

CONCLUSIONS

We related systematic variations in interparticle attraction to the extrusion and centrifugal casting behavior of flocculated alumina suspensions. Samples prepared by salt flocculation and with surface-adsorbed fatty acids exhibited a shift from high to low compressibility at a maximum density that increased with decreasing interparticle attraction. In contrast, strongly-flocculated systems that were prepared without short-range repulsion near the isoelectric point were highly compressible over a broad range of effective stresses. Centrifuged samples that were prepared using these strongly-flocculated slurries exhibited nonuniform packing densities as a result of nonuniform consolidation stresses and the compressible nature of these systems. In contrast, packing densities were highly uniform after decreasing the interparticle attraction by the other methods used.

As reduced interparticle attraction rendered upward shifts in the maximum density at the transition from high to low compressibility, we hypothesized that this reduced attraction may also increase the threshold density at which ductile-to-brittle transitions occur during plastic shear. Samples prepared with oleic acid were highly plastic and easily extrudable at solids densities of up to 59 vol%. In contrast, salt-flocculated samples at 55 vol% density extruded at creeping flow rates that were approximately independent of the applied stress. Hand-molding tests revealed that salt-flocculated samples were creep-sensitive at low stresses, and higher stresses produced a transition to stiff and brittle behavior. Results suggested that particle rearrangement during plastic flow is a rate-limiting process, with an average relaxation time that is reduced with reduced interparticle attraction.

ACKNOWLEDGMENTS

This research is supported by the Office of Basic Energy Sciences at the U.S. Department of Energy through a subcontract by Pacific Northwest Laboratory under Contract No. 063961-A-F1 and by the Foundation Blancheflor Boncampi-Ludovisi née Bildt. The authors wish to thank Lilian Thompson, Greg Exarhos, and Gary McVay for their assistance.

REFERENCES

1. F. M. Tiller and C. D. Tsai, "Theory of Filtration of Ceramics: 1, Slip Casting," *J. Am. Ceram. Soc.* **69** (12) 882-87 (1986).

2. C. H. Schilling, W-H. Shih, and I. A. Aksay, "Advances in the Drained Shaping of Ceramics," in *Ceramic Powder Science IV, Vol. 22*, editors S. Hirano, G. L. Messing, and H. Hausner (American Ceramic Society, Westerville, Ohio, 1991), pp. 307-21.

3. C. H. Schilling, "Plastic Shaping of Colloidal Ceramics," Ph.D. Thesis, University of Washington, Seattle (1992).

4. L. Bergström, C. H. Schilling, and I. A. Aksay, "Consolidation Behavior of Flocculated Alumina Suspensions," *J. Am. Ceram. Soc.* **74** (12) (1992) in press.

5. R. Buscall, "The Sedimentation of Concentrated Colloidal Suspensions," *Colloids and Surfaces* **43**, 33-53 (1990).

6. F. F. Lange, "Powder Processing Science and Technology for Increased Reliability," *J. Am. Ceram. Soc.* **72** (1) 3-15 (1989).

7. C. H Schilling, J. J. Lannutti, W.-H. Shih, and I. A. Aksay, "Stress-Density Variations in Alumina Sediments: Effects of Polymer Chemistry," in *Mechanical Properties of Porous and Cellular Materials, M.R.S. Symp. Proc. Vol. 207*, editors D. J. Green, L. J. Gibson, and K. Sieradski (Materials Research Society, Pittsburgh, PA, 1991), pp. 151-56.

8. V. K. Pujari, "Effect of Powder Characteristics on Compounding and Green Microstructure in the Injection Molding Process," *J. Am. Ceram. Soc.* **72** (10) 1981-1984 (1989).

9. G. C. Stangle and I. A. Aksay, "Simultaneous Momentum, Heat and Mass Transfer with Chemical Reaction in a Disordered Porous Medium: Application to Binder Removal from a Ceramic Green Body," *Chem. Eng. Sci.* **45** (7) 1719-1731 (1990).

10. A. M. Homola, J. N. Israelachvili, M. L. Gee, and P. M. McGuiggan, "Measurements of and Relation Between the Adhesion and Friction of Two Surfaces Separated by Molecularly Thin Liquid Films," *J. Tribol.* **111**, 675 (1989).

     Characterization Techniques for the Solid-Solution Interface

11. B. V. Velamakanni, J. C. Chang, F. F. Lange, and D. S. Pearson, "New Method for Efficient Colloidal Particle Packing via Modulation of Repulsive Lubricating Hydration Forces," *Langmuir* **6**, 1323-1325 (1990).

12. L. Bergström, "Sedimentation of Flocculated Alumina Suspensions: Gamma-Ray Measurements and Comparison with Model Predictions," *J. Chem. Soc. Faraday Trans.*, **88** (1992), in press.

13. G. Frens, "Interacting Particles in Contact," *Faraday Disc. Chem. Soc.*, **90**, 143-51 (1990).

14. R. G. Horn, "Surface Forces and Their Action in Ceramic Materials," *J. Am. Ceram. Soc.* **73** (5) 1117-1135 (1990).

15. C. H. Schilling, G. L. Graff, W. D. Samuels, and I. A. Aksay, "Gamma-Ray Densitometry: Non-Destructive Analysis of Density Evolution During Ceramic Powder Processing," in *Atomic and Molecular Processing of Electronic and Ceramic Materials: Preparation, Characterization, and Properties*, editors I. A. Aksay, G. L. McVay, T. G. Stoebe, and J. Wager (Materials Research Society, Pittsburgh, PA, 1987), pp. 239-51.

16. F. F. Lange and K. T. Miller, "Pressure Filtration: Consolidation Kinetics and Mechanics," *Am. Ceram. Soc. Bull.* **66** (10) 1498-1504 (1987).

17. H. P. Grace, "Resistance and Compressibility of Filter Cakes," *Chem. Eng. Prog.* **49** (6) 303-18 (1953).

18. T. J. Fennelly and J. S. Reed, "Compression Permeability of Alumina Cakes Formed by Pressure Slip Casting," *J. Am. Ceram. Soc.* **55** (8) 381-83 (1972).

19. C. P. Wroth and G. T. Houlsby, "Applications of Soil Mechanics Theory to Ceramics Processing," in *Ultrastructure Processing of Ceramics, Glasses, and Composites,* edited by L. L. Hench and D. R. Ulrich (Wiley, New York, 1984), pp. 448-63.

20. R. E. Reed-Hill, *Physical Metallurgy Principles* (PWS, Boston, 1973).

21. L. H. Sperling, *Physical Polymer Science* (Wiley, New York 1985).

22. M. R. Kuhn and J. K. Mitchell, "The Modelling of Soil Creep with the Discreet Element Method," *Proc. of the First U.S. Conference on Discreet Element Methods*, edited by G. G. W. Mustoe, M. Hendricksen, and H. P. Huttelmaier (CSM Press, Golden, Colorado, 1989).

MANIPULATING SOLID-SURFACE PROPERTIES WITH POLYMERIC AGENTS

Jean-François Argillier and Matthew Tirrell, University of Minnesota, Department of Chemical Engineering and Materials Science, Minneapolis, MN 55455

ABSTRACT

Measurements have been made of the forces of interaction between pairs of adsorbed layers of diblock copolymers. These have been made on layers of 2-vinylpyridine-styrene and 2-vinylpyridine-isoprene copolymers. From toluene solution, the vinylpyridine block adsorbs strongly and the other blocks adsorb negligibly. This paper discusses the adsorption process itself and ways in which the interactions between two such layers can be tailored for particular purposes.

INTRODUCTION

Fabrication of advanced ceramics involves the synthesis and processing of submicron ceramic particles of a wide variety of compositions, shapes and sizes[1,2]. The main deficiency of conventional ceramics that modern ceramics processing is trying to overcome is not an inherent lack of strength or stiffness but rather the undependable brittleness due to the ease with which cracks start and propagate. Structures that are homogeneous at a fine scale are therefore desired.

Colloidal processing, that is, handling the uniform size particles in a liquid dispersion, offers the possibility of minimizing undesirable heterogeneities, either by removing large aggregates in a separate step or by effectively complete stabilization of the primary particles. Subsequent consolidation by slip casting or pressure filtration, tape casting or injection molding then produces quite uniform microstructures and high packing densities with flaw sizes reduced to the size of the submicron particles themselves.

Manipulation of the interparticle potential through the use of polymeric agents enters into the fabrication of advanced ceramics in the following ways:

Characterization Techniques for the Solid-Solution Interface   53

--creation of stable dispersions that can be readily compressed to dense ordered structures;

--determination of the minimum amount of polymeric agent necessary to produce a suitable potential in order to minimize the amount of organic material that must eventually be burned out of (and possibly contaminate) the ceramic material;

--using the interparticle potential to manipulate the mechanical and rheological properties of the materials during processing.

Block copolymers in solution afford the opportunity to create unique adsorbed layers via selective adsorption from a selective solvent. This process uses the amphiphilic character of block copolymers in that it can be arranged that one block adsorbs strongly on a solid surface and avoids the solvent, while the other block, nonadsorptive on its own, swells into the good solvent surrounding medium but remains tethered to the surface by the adsorbing block.[3-5] This provides a means to modify the interactions of solid surfaces immersed in solvents. One block of the copolymer anchors firmly while the other extends out and produces the desired interaction with the environment.

Besides processes for ultrafine ceramics, this self-assembly process has proven to be very useful in technology to modify the surfaces of colloidal particles that are used in coatings, paints, xerography, photography[6]. The scientific questions that are posed in this situation concern the nature of the configurations adopted by the macromolecules in this environment and how peculiarities in this configuration may reflect themselves in the nature of the interactions that these layers produce with the surroundings.

The following facts appear to be established concerning the molecular arrangements in these layers. In the interesting case that the anchoring energy, $\varepsilon_A N_A$, is large, where ε_A is the binding energy per anchoring segment and N_A is the number of anchoring segments, the number of chains adsorbed per unit area, σa^{-2}, (with a being the size of a monomer unit) can be sufficiently high that it exceeds the threshold for overlap of the nonadsorbing chains (the "buoys"). This criterion for overlap of the buoys can be expressed as:[7]

$$\sigma_{ol} = N_B^{-6/5} \tag{1}$$

where N_B is the number of segments in the buoy block. Only if $N_A \gg N_B$, or if the anchoring energy is very weak, is this criterion not met.[5] Under these circumstances, the buoy chains form a polymer "brush"[7-10], in which the chains extend well beyond their solution dimensions in the direction normal to the adsorbing plane (Figure 1).

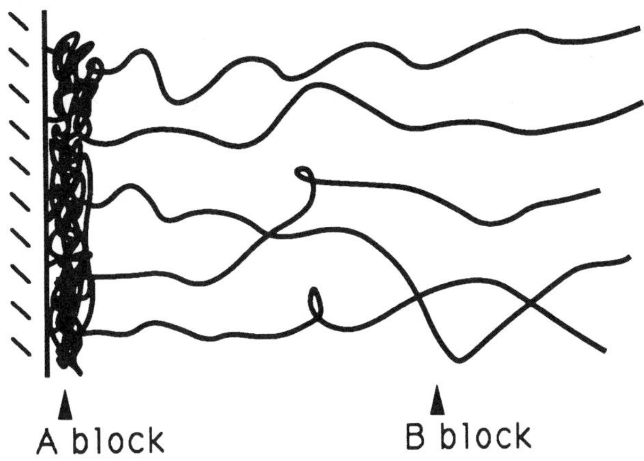

Figure 1 Adsorption of A-B diblock copolymers from a selective solvent. The poorly solvated A block adsorbs on the surface and the well solvated B block forms a polymer brush.

That a polymer brush should be extended can be understood from a Flory-type argument, balancing osmotic swelling of the dense brush with elastic resistance to swelling. In terms of the free energy per chain in the brush, F/kT, this argument can be written:

$$F/kT = L^2/2Na^2 + vN(Na\sigma/L) \qquad (2)$$

where v is a dimensionless excluded volume ($\sim 1 - 2\chi$) and the quantity in parentheses is the volume fraction of polymer segments in a uniformly stretched brush of height L. Minimization of this energy with respect to L gives:

$$L = Na(v\sigma)^{1/3} \qquad (3)$$

The linear dependence of L on N signals that chains in a brush have characteristic dimensions that scale more like chains standing on end than random coils, though, as we shall demonstrate, typical values of σ that are achieved by the examples of selective adsorption that have been tried so far are of order 0.01, so that, L << Na.

This equation has been tested and appears to agree reasonably well with the length scale of brushes as measured both by surface forces measurement and by neutron scattering.[11,12] However, this should lead neither to the conclusion that the profile of segments is consistent with uniform stretching nor that Equation 2 is an accurate one for the stretching energy. It has been clearly demonstrated by self-consistent field calculations[13a] and by molecular dynamics simulations[13b] that the stretching of the chains is not uniform (decreasing with distance from the surface) and that chain ends, implicitly assumed in Equation 2 to be all at the tip of the brush, in fact, distribute themselves throughout the depth of the brush. The resulting profile of segments is more like parabolic[13] than uniform, though Equation 3 remains a good quantitative measure of the average brush height.

The reasons why Equation 2 does not give a good measure of the stretching energy, even though Equation 3 is adequate, have been discussed by de Gennes[14] and by Freed[15]. The principal reason is that both terms in Equation 2 are overestimates that cancel in going to Equation 3. To get a more accurate calculation of the stretching energy, both terms must be modified to account for correlations present in the strong excluded volume environment in good solvent. For example, the osmotic pressure energy should be proportional to $(N\sigma a/L)^{5/4}$ [16], instead of linear, since correlations reduce the number of binary contacts between segments. Using this to estimate the energy by substituting Equation 3 into the osmotic term of Equation 2 modified in this way gives a stretching energy per <u>layer</u> (= energy per chain times σ):

$$F_{layer}/kT = Nv^{7/12}\sigma^{11/6} \qquad (4)$$

This equation represents the penalty that a chain must pay to adsorb in a high density brush. This will be the basis for the analysis of the adsorption data that we present in this paper.

The other aspect of polymer brushes that this paper will discuss is the means by which the profile of segments extending away from the surface can be tailored. In a sense, the nonuniform stretching recognized by the Milner,Witten and Cates[13] theory, though clearly correct, is a disappointment to the extent that layers with all of the chain ends at the periphery may have special properties or utility. For example, if a synthetic reaction were designed to place a particular functional group at the ends of the chains this could be used to create a brush with an array of "stickers" at the tips of the bristles. The analysis of the polydisperse brush[17] shows that ends of longer chains are segregated from the shorter, denser brush for essentially the same reason that brushes are inpenetrable to other brushes or solutions of equal or lower segment density. It is more favorable for the extra chain length to reside above the dense brush than to penetrate it and cause the dense brush to stretch still more.

     Characterization Techniques for the Solid-Solution Interface

One means to tailor the profile of segments in a brush is to use a specified distribution of chain lengths in the assembly of the brush. A few longer bristles on the brush bearing the stickers may recover some of the desirable feature of the hypothetical uniformly stretched brush. Two questions (at least) arise in this connection. One is whether or not the competitive adsorption process strongly favors one block copolymer over another. We have and present evidence here for what the thermodynamically favored tendencies are. Kinetic favorization is also possible, even likely, though it is unstudied as yet in competitive adsorption. We report here on our first measurements on the self-assembly and forces exerted between bimodal brushes. These have been made on layers of 2-vinylpyridine-styrene (PVP-PS) and 2-vinylpyridine-isoprene (PVP-PI) copolymers. From toluene solution, the vinylpyridine block adsorbs strongly and the other blocks adsorb negligibly.[1]

EXPERIMENTAL

The methods we have used to determine the adsorbed amounts of PVP-PS and PVP-PI block copolymers adsorbed on mica from toluene have been described in detail in another publication.[5] We discuss here only the most salient features, as well as the aspects of the method necessary to study bimodal brushes. All of the polymers were prepared by anionic polymerization and had molecular weights distributions with a polydispersity of less than 1.1. Table I lists all of the block copolymers on which we have measured adsorbed amounts. Surface forces measurements have been made on a subset of these. Several of these samples, those including a prefix t- in their sample designations, have been synthesized incorporating a radioactive label. This was done by terminating the living polymerization with tritiated methyliodide (CT_3I). Scintillation counting on the carefully and completely collected combustion products from incineration of the adsorbed layer is then an accurate means of determining the weight of polymer adosorbed per unit area.

However, the method we have developed is more versatile than scintillation counting alone. X-ray photoelectron spectroscopy (XPS) is a very accurate method for the determination of the relative adsorbed amounts in two adsorbed layers. It does so by measuring the attenuation of the signal from a particular photoelectron (Si_{2p} in the case of mica or silica) in the substrate. We have shown that for a series of chemically similar materials such as the range of composition represented in Table I, the escape depth for photoelectrons varies by only about 1Å, so that the normalized signal from the attenuated photoelectron is directly proportional to the surface coverage. If the coverage of one or more of the polymers is known, the relative measure provided by XPS can be calibrated into a absolute method. Calibration with the independently determined adsorbed amounts of the tritiated polymers enables this to be done. In our hands the precision of determination of the adsorbed amounts by this method is approximately ± 5%.

Scintillation counting also permits the determination of the composition in a mixed adsorbed layer containing two species of chains. In this case, the total adsorbed amount is determined by XPS and the fraction of a tritiated species in the adsorbed layer is determined by scintillation counting. If the internal copolymer compositions of the two species used to assemble the bimodal layer is sufficiently different, we have found that we can also determine the composition of the adsorbed layer directly from XPS by measuring the carbon to nitrogen elemental composition ratio of the adsorbed layer and using the known elemental compositions of the copolymers.

The methods used in our laboratory for surface forces measurement have also been described in detail previously[11]. We refer the interested reader there for details. The adsorption is done in all cases inside the apparatus under conditions otherwise identical to those under which the layers on which surface coverage was measured were assembled. All force curves reported are believed to represent equilibrium forces exerted at the indicated distances, with approaching force curves equivalent to separating force curves within experimental accuracy. As is convenient and conventional, all force versus distance data are plotted as F/R vs. D, where R is the measured local radius of curvature of the mica sheets in the force-measuring zone, thereby convertingthe data into a quantity which is geometry-independent and proportional to the energy per unit area of interaction between the polymer-bearing surfaces.

RESULTS AND DISCUSSION

<u>Adsorption measurement</u>

Table I gives the measured adsorbed amounts for the range of block copolymers that we have studied. The data are given both as weight per unit area, s, and number of chains per unit area, σa^{-2}. The fourth column of Table I gives the values of σ_{ol} for each of the molecular weights of buoy block, calculated using Equation 1. All but the largest PVP molecular weights give $\sigma > \sigma_{ol}$, so the vast majority of these layers have self-assembled into polymer brushes. An interesting observation is that the weight per unit area does not vary greatly over the range of copolymers we have studied.

Figure 2 displays these data as a 3-D perspective plot where σa^{-2} is plotted versus both molecular weights of the two blocks. It is difficult to see clear trends in these data except that a fair generalization appears to be that the number of chains adsorbed per unit area decreases with increasing molecular weight of either of the two blocks. In order to derive more information from these data, we have attempted to find a combination of the two block molecular weights that would permit the collapse of the data as a function of both molecular weights to be represented as a single function of a combined

Table 1 Summary of adsorbed amounts for PVP-PS and PVP-PI diblock copolymers. The final column is the surface density (σ_{ol}) above which the PS buoys will overlap in the adsorbed layers.

Sample Mw of PVP-PS (10^3 g/mol)	Adsorbed Amount s(mg/m^2)	Surface Density σ(m^{-2}x10^{-16})	$1/(\pi R_{PS}{}^2)$ σ_{ol}(m^{-2}x10^{-16})
3-36	2.16	3.32	0.84
9-36	2.12	2.84	0.84
16-31	1.79	2.30	1.00
18-36	1.83	2.05	0.84
15.92	2.02	1.13	0.27
15-152	1.97	0.71	0.15
31-31	1.66	1.61	1.00
36-36	1.49	1.25	0.84
31-92	1.82	0.89	0.27
30-152	1.89	0.63	0.15
62-31	1.36	0.88	1.00
61-92	1.65	0.65	0.28
61-152	2.05	0.58	0.15
72-36	1.67	0.93	0.84
124-31	1.57	0.61	1.00
t52-63	1.61	0.84	0.43
t9-32	1.95	2.90	0.98
t124-60	1.13	0.37	0.45
Sample PVP-PI			
26-50		1.25	0.39
30-217		0.34	0.067
38-69		1.00	0.26
69-39		0.87	0.52

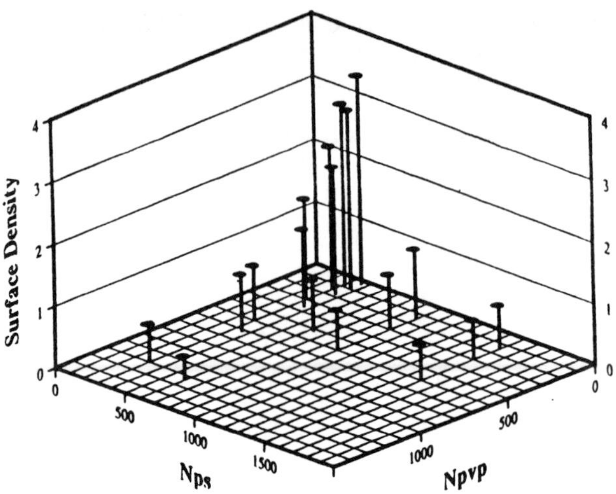

Figure 2   Measured surface densities $\sigma(m^{-2}x10^{-16})$ of Table 1 as a function of both the PS (N_{PS}) and PVP (N_{PVP}) degrees of polymerization. Vertical lines are used to indicate heights of data ponts above and their coordinates in N_{PS} - N_{PVP} plane.

molecular weight parameter. Using the simple ratio of the two block molecular weights does not work in a particularly impressive manner. What does work very well, as shown in Figure 3, is to normalize σ by σ_{ol}, $\equiv \sigma^*$, and plot it as a function of what we shall term the "solvent-induced asymmetry" of the copolymer, $\beta \equiv N_B^{6/5}/N_A^{2/3}$, which is the square of the ratio of the swollen radius of the buoy block to the collapsed radius of the anchor block. This parameter reflects both the molecular weight and solvent quality asymmetry of the block copolymer. σ^* is a measure of how crowded the buoys are.

Clearly the normalized number density of adsorbed chains is a unique function of this asymmetry parameter to a good degree of approximation. Furthermore, for all but the copolymers of highest asymmetry, σ^* appears to be a power law function of β. That this is reasonable can be argued from the theory of Marques, et al[18], on the selective adsorption of block copolymers from selective solvents. They have analyzed the physics of this adsorption thoroughly and we present here a drastically simplified extract and minor modification of their theory.

                 Characterization Techniques for the Solid-Solution Interface

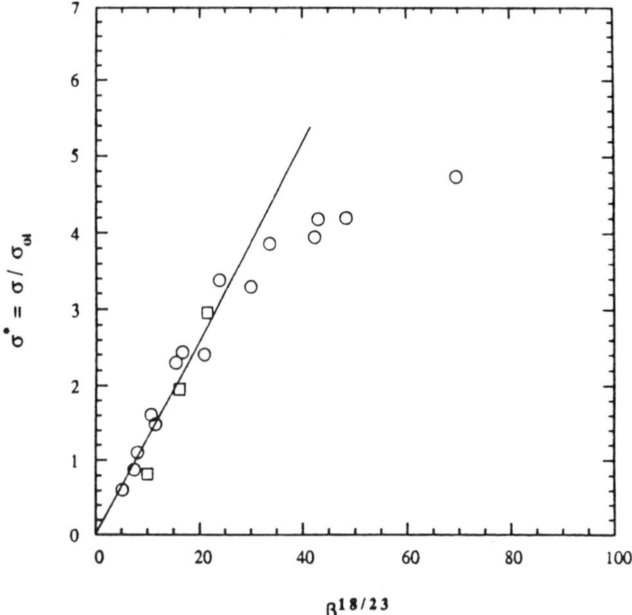

Figure 3 Normalized surface density σ* versus ß18/23 as measured by scintillation ocunting from tritium labeled adsorbed layers (squares) and XPS (circles). The solid line represents a least squares fit to the linear portion of the data (excluding five data points at highest ß values) as predicted from Equation 6.

We imagine the assembly of the layer from solution to be governed by a balance of only two energies. In this situation, the relevant energies are the adsorption energy of the anchor (favorable for adsorption) and the stretching energy of the brush (unfavorable for adsorption). Marques, et al, propose that the adsorption of the anchoring block takes place as though droplets of the material composing the anchor were wetting the surface and carrying along the buoy chains. This leads to an expression for the energy of the anchor layer that takes the form of a van der Waals energy, varying like the inverse square of the thickness of the layer, δ, and containing an effective Hamaker constant that is a composite parameter of the relevant material Hamaker constants. Balancing this against the stretching energy of a layer containing σ chains (Equation 4, assuming an athermal solvent for the buoy chains, so that $\chi = 0$ or $v = 1$) gives, for the assembly of the layer:

$$F/kT = A/12kT\pi\delta^2 + N_B\sigma^{11/6} \qquad (5)$$

The wetting anchor layer is a dense layer with a polymer volume fraction, $\phi = N_A a\sigma/\delta$, near unity, so $\delta \approx N_A a\sigma$. Inserting this into Equation 5, and minimizing with respect to σ, then recasting the result as σ^* as a function of β, gives:

$$\sigma^* \approx (A/kT)^{6/23}\beta^{18/23} \qquad (6)$$

For all but the copolymers of highest asymmetry, this equation is in good agreement with the data of Figure 3.

Clearly this curve must actually have a maximum at high asymmetry since this corresponds to negligible anchor block size and therefore to no adsorption ($\sigma^* = 0$). The lattice model theory of Evers, et al predicts this quantitatively[19] and is, on preliminary examination, also able to explain some aspects of our data. This warrants a more thorough examination. We know empirically from our data that the deviation from power law behavior occurs at a number density of chains where the number of segments in the PVP blocks of those chains are in insufficient number to cover the surface completely, making the idea of a continuous wetting anchor layer untenable for these molecules.

<u>Bimodal brushes</u>

With the understanding we have developed for the unimolecular adsorption process, we can already suggest some expections for equilibrium aspects of the assembly of bimodal brushes. For pairs of polymers such as the 61-92 and 61-152 samples of Table I, where the anchoring block molecular weight is the same and the adsorbed number densities in the pure layers are nearly the same, the major difference is in the stretching energy of the brush. Adsorption from a 50% mole fraction solution should produce a layer where the strething energy of the shorter component is largely unaffected where that of the longer component is favorably affected, relative to their respective pure component layers. Therefore, the total number density of chains should go up in the mixed layer.

In practice, the only bimodal brush layer on which we have data thus far is that composed of the samples 9-36 and 15-152 from Table I, where the measured surface densities of the pure component layers are reported. This is not an example of the idealized case described above. Experimentally, we find that adsorption from a 50 wt % solution of these two species at a total polymer concentration of 100 ppm produces a layer that is 42 wt % of the 15-152 polymer, with a total number density, $\sigma_T a^{-2} \approx 1.5 \times 10^{16}$ chains/m^2. This is less than the arithmetic mean of the number densities of the pure component layers. Table II gives some other of the important characteristics of the mixed adsorbed layer. $\sigma_m a^{-2}$ is the number per unit area of that species in the mixed adsorbed layer.

   Characterization Techniques for the Solid-Solution Interface

Table II
Characteristics of Mixed Bimodal Adsorbed Layers

Polymer component in the layer	$\sigma_m a^{-2}$ (10^{16} chains/m^2)	$\sigma_{ol} a^{-2}$ (10^{16} chains/m^2)
15-152	0.24	0.15
9-36	1.24	0.84

Both species in the mixed layer exceed their overlap densities, which is a sufficient condition for the remaining chain lengths of the long species to continue to overlap among themselves after emanating from the short brush. It is possible to adapt the arguments of Alexander[7] and de Gennes[8] to this type of bimodal brush[20]. Before describing that analysis it is worth noting that all bimodal brushes are not necessarily of this category. It is also possible to make bimodal brushes with nonoverlapping in either the short or the long component, producing layers that are characterized by "underbrush" or "mop" structures, respectively. We shall deal only with the case where the long chains remain overlapped in the mixed layer, a structure we will call a "double brush", for simplicity. The short chains are characterized by a polymerization degree N_S and an adsorbed surface density σ_S. The long chains are characterized by a polymerization degree N_L and an adsorbed surface density σ_L. The total surface density σ_T is given by $\sigma_T = \sigma_L + \sigma_S$. The bimodal layer is composed of an inner and an outer brush (figure 4)

The inner brush is constituted of the short chains and of N_S segments of the long chains. As the solution is a semi-dilute solution the osmotic pressure should de determined by the local monomer volume fractions $c_T = c_S + c_L$, where c_S, c_L are respectively the volume fractions occupied by monomers of the short and long chains. Following Alexander[7] and de Gennes[8] approach, these N_S segments can be subdivided into blobs of radius d_T, where $d_T = a\sigma_T^{-1/2}$. Each blob contains $g_i = (d_T/a)^{5/3}$ monomers and the thickness of the inner brush scales as the number of blobs times the size of the blobs, that is

$$L_i = d_T N_S/g_i = a N_S \sigma_T^{1/3} \qquad (7)$$

The outer brush is composed of chain lengths of N_L-N_S segments arranged in $(N_L-N_S)/g_L$ blobs of size d_L. As $g_L = (d_L/a)^{5/3}$, its thickness L_o is then given by:

$$L_o = a(N_L-N_S)\sigma_L^{1/3} \qquad (8)$$

Calling α the asymetryc ratio of the polymeric species ($\alpha > 1$),

$$\alpha = N_L/N_S \tag{9}$$

and λ the fraction of long chains grafted on the surface ($0 < \lambda < 1$),

$$\lambda = \sigma_L/\sigma_T \tag{10}$$

the total thickness of the bimodal brush $L_m = L_i + L_o$ is given by:

$$L_m = aN_S\sigma_T^{1/3}(1 + (\alpha-1)\lambda^{1/3}) \tag{11}$$

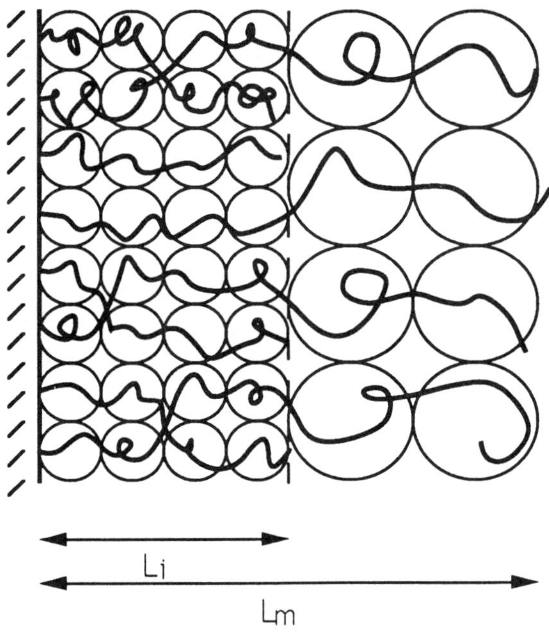

Figure 4   Description of a bimodal grafted brush. The thickness of the inner brush is given by $L_i = aN_S\sigma_T^{1/3}$, whereas the total thickness of the brush is given by $L_m = aN_S\sigma_T^{1/3}(1 + (\alpha-1)\lambda^{1/3})$.

Figure 5 presents data on measurements of force versus distance profiles between brushes assembled from the components of Table II. Data are shown for the two pure layers and for the mixed layer for which the relevant characteristics are given in Table II.

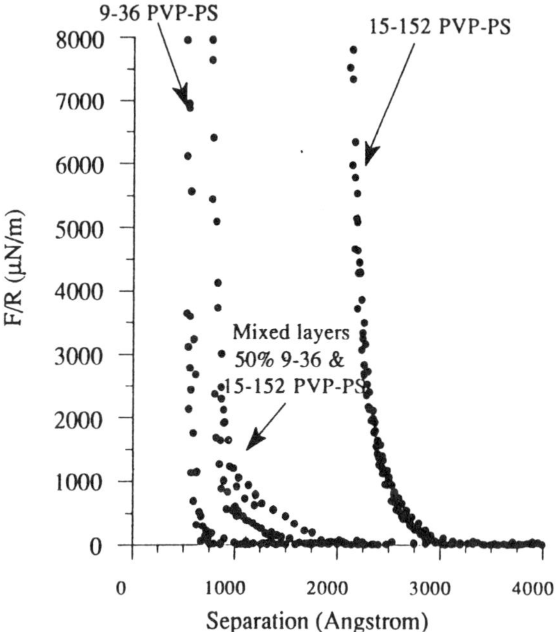

Figure 5 Force vs. distance profiles for pure and mixed layers of PVP-PS block copolymers 9-36 and 15-152.

A complete quantitative analysis of these data is not yet complete but one can extract characteristic length scales from Figure 5 and compare them with expectations based on Equations 3 and 7. For the surface densities reported in Table I, Equation 3 predicts that the ratio of layer thicknesses, which we will take (somewhat arbitrarily) as half the distance at which the force rises to a value experimentally distinguishable from zero ($\approx 50\mu N/m$), $L_L/L_S = (N_L/N_S)(\sigma_L/\sigma_S)^{1/3}$, which for the values given in Table I on these materials gives: $L_L/L_S = 2.83$. Experimentally, the measured ranges of the forces have the ratio $L_L/L_S \approx 2950/900 = 3.27$, which is about 16% above the prediction. With this information on the pure layers as background it is of more interest to examine the range of the forces in the self-assembled mixed layers. From Equations 3 and 7, inserting the data from Tables I and II, the prediction for the ration of the mixed layer thickness to the pure short layer thickness is

$L_m/L_S = 2.26$. Experimentally, this ratio is $L_m/L_S \approx 1600/900 = 1.77$, which is about 21% below the prediction. Bear in mind that these predictions used experimental data on surface densities and so are not liable to errors in assumptions about the assembly of the layer (though of course they are liable to errors in the measurements of surfaces densities).

It appears reasonable to say at this point that the Alexander-de Gennes type of model for the bimodal brush gives a reasonable account of their characteristic length scales at a level of about 20% accuracy. Further work is required to strengthen this conclusion and to explore other interesting and important potential modifications of the profiles of polymer brushes.

In summary, it has been shown that self-assembly by selective adsorption of diblock copolymers from a selective solvent is an expedient means to produce a polymer brush. We have shown how the stretching engendered by the construction of the brush is an important determinant of the assembly process.Construction of bimodal double brushes and other mixed configurations will prove useful in tailoring profiles of segments.

REFERENCES

1. D.R. Ulrich, Chemical and Engineering News, January 1, 1990.
2. L. L. Hench and D. R. Ulrich, eds., "The Science of Ceramic Chemical Processing", Wiley, N.Y., 1986.
3. G. Hadziioannou, S. Patel, S. Granick and M. Tirrell, "Forces between surfaces of block copolymers adsorbed on mica", *J. Am. Chem. Soc.*, **108**, 2869 (1986); M. Tirrell, S. Patel and G. Hadziioannou, "Polymeric Amphiphiles at Solid-Fluid Interfaces: Forces between layers of adsorbed block copolymers", *Proc. Natl. Acad. Sci. USA*, **84**, 4725 (1987).
4. H.J. Taunton, C. Toprakcioglu and J. Klein," Direct measurement of the interaction between mica surfaces with adsorbed diblock copolymer in a good solvent", *Macromolecules*, **21**, 3333 (1988).
5. E. E. Parsonage, M. Tirrell, H. Watanabe and R. G. Nuzzo, " Adsorption of poly2vinyl pyridine-polystyrene block copolymers from toluene solution", *Macromolecules*, **23**, in press (1990).
6. W. B. Russel, D. A. Saville and W. R. Schowalter, "Colloidal Dispersions", Cambridge University Press, Cambridge, 1989.
7. S. Alexander, "Adsorption of chain molecules with a polar head: A scaling description", *J. Phys. (Paris)*, **38**, 983 (1977).
8. P.-G. de Gennes, "Conformations of polymers attached to an interface", *Macromolecules*, **13**, 1069 (1980).
9. S. Patel, M. Tirrell and G. Hadziioannou, "A simple model for forces between surfaces bearing grafted polymers applied to data on adsorbed block copolymers", *Colloids and Surfaces*, **31**, 215 (1988).

     Characterization Techniques for the Solid-Solution Interface

10. H.J. Taunton, C. Toprakcioglu, L.J. Fetters and J. Klein, "Interactions between surfaces bearing end-adsorbed chains in a good solvent", *Macromolecules,* **23**, 571 (1990).

11. S. Patel and M. Tirrell, "Measurement of forces between surfaces in polymer fluids", *Ann. Rev. Phys. Chem.*, **40**, 597 (1989).

12. Philippe Auroy, "Polymères Greffés à l'Interface Solide-Liquide", Thèse, Université de Paris-Sud, 1990.

13. a) S. J. Hirz, "Modeling of Interactions between Adsorbed Block Copolymers", M.S. Thesis, University of Minnesota, 1986; S. T. Milner, T. A. Witten and M. E. Cates, "Theory of the grafted polymer brush", *Macromolecules*, **21**, 2610 (1988). b) M. Murat and G. S. Grest ," Stucture of a grafted polymer brush: A molecular dynamics simulation",*Macromolecules*, **22**, 4054 (1989).

14. P.-G. de Gennes, "Scaling Concepts in Polymer Physics", Cornell University Press, Ithaca, 1979, Chapter 2.

15. K.F. Freed, "Renormalization Group Theory of Macromolecules", Wiley, N.Y., 1987, pgs 66-70.

16. I. Noda, N. Kato, T. Kitano and M. Nagasawa," Thermodynamic properties of moderatively concentrated solutions of linear polymers", *Macromolecules*, **14**, 668 (1981).

17. S. T. Milner, T. A. Witten and M. E. Cates, "Effects of polydispersity in the end-grafted polymer brush", *Macromolecules*, **22**, 853 (1989).

18. C. M. Marques, L. Leibler and J.F. Joanny, "Adsorption of block copolymers in selective solvents", *Macromolecules*, **21**, 1051 (1988).

19. O. A. Evers, J. M. H. M. Scheutjens and G. J. Fleer, "Statistical thermodynamics of block copolymer adsorption.Part 2. Effect of chain composition on the adsorbed amount and layer thickness",*J. Chem. Soc. Faraday Trans.*, **86**, 1333 (1990).

20. J.-F. Argillier, M. Tirrell, "Tailored grafted polymer layer on a solid-liquid interface", *Macromolecules*, to be published.

A GENERALIZED PROGRAM TO CALCULATE INTERPARTICLE INTERACTIONS IN A VARIETY OF SUSPENSION CONDITIONS

James H. Adair and Richard V. Linhart
Department of Materials Science and Engineering
University of Florida, Gainesville, FL 32611.

INTRODUCTION

It has been shown by numerous investigations [1-5] that the theory articulated by Deryagin and Landau [6] and Verwey and Overbeek [7]* predicting electrostatic stability for colloidal suspensions is generally valid. Although there have been many excellent reviews given on these theories [1-5], they have dealt with the limited, simplified cases of symmetrical electrolytes (e.g., NaCl or $CaSO_4$). This assumption allowed the original developers of the theory [6,7] to derive differential equations describing the relationship of electric potential as a function of perpendicular distance from the solid surface in simplified forms for which analytical solutions or tables of numerical solutions were available [8]. However, for solutions of electrolytes which are asymmetrical or limited to a discrete number of ionic species, it is necessary to use more general mathematical relationships describing the undisturbed diffuse double layer as well as interacting double layers. Most systems of interest, particularly ceramic materials and biologically relevant systems, have complex ionic equilibria associated with the solid phase. Therefore, the goal of the work presented in this document is to develop and present an approach to predicting the colloid stability in suspensions containing complex ionic equilibria.

---

* $\underline{D}$eryagin and $\underline{L}$andau [6] and $\underline{V}$erwey and $\underline{O}$verbeek [7], for whom the theory is designated DLVO theory

THE UNDISTURBED DIFFUSE DOUBLE LAYER

Grahame [9] has given generalized equations describing the mathematical relationship between potential and distance for the Gouy-Chapman layer. This relationship was obtained based upon the premise that the Poisson equation may be utilized to describe the divergence of the slope of potential versus distance as a function of the space charge density of the counter ions, and that the Boltzmann equation describes the concentration gradient of ions as a function of electric work only. The assumptions inherent to a development of the Poisson-Boltzmann equation are:

1. the solid is a flat semi-infinite plate, such that the double layer is flat;
2. the dielectric constant near the surface may be taken to be that of the bulk solution
3. the ions are considered as point charges, where their volume is taken to be zero, and;
4. the work ($W(x)$) done in transferring the ions from the bulk solution to the double layer is only performed against Coulombic forces where $W(x) = z_i e\psi(x)$, where z_i is the valence of the ion, e is the charge on an electron, and $\psi(x)$ is the potential in volts at a distance, x.

These give the Poisson-Boltzmann equation

$$\frac{d^2\psi(x)}{dx^2} = -\frac{4\pi e}{DD_o}\sum n_i v_i \exp[-\frac{z_i e\psi(x)}{kT}] \tag{1}$$

where $\psi(x)$ is the potential in volts at a distance, x, in centimeters, from the Stern plane, e is the charge of a proton in Coulombs, D is the relative dielectric constant of the solution, D_o is 4π times the permittivity of free space taken to be equal to 1.112×10^{-12} C V^{-1} cm^{-1}, n_i is the concentration in ions cm^{-3}, of the ith ionic species, z_i is the corresponding valence, k is the Boltzmann constant in J °K^{-1} ion^{-1}, and T is the absolute temperature of the solution. First integration of the Poisson-Boltzmann equation under the boundary conditions that as x approaches infinity, $\psi(x)$ approaches zero and $d\psi(x)/dx$ approaches zero, gives

   Characterization Techniques for the Solid-Solution Interface

$$\frac{d\psi(x)}{dx} = \pm\sqrt{\frac{8\pi kT}{DD_o}}\sqrt{\sum n_i(\exp(-\frac{z_i e\psi(x)}{kT})-1)} \qquad (2)$$

Execution of the second integration of the Poisson-Boltzmann equation yielding potential as a function of distance from the Stern plane will be reserved for the next section which discusses the situation with overlapping double layers. Surface charge (σ_o), related to Equation 2 by $-4\pi/DD_o$, is given by

$$\sigma_o = \pm\sqrt{\frac{DD_o kT}{2\pi}}\sqrt{\sum n_i(\exp(-\frac{z_i e\psi_o}{kT})-1)} \qquad (3)$$

where σ_o is the surface charge in $C{\cdot}cm^{-2}$ and ψ_o is the potential at the surface. When a Stern layer is present, ψ_o may be replaced by the Stern potential (ψ_δ) to yield the charge of the Gouy-Chapman layer (σ_{GC}).

OVERLAPPING DOUBLE LAYERS

When two identical particles approach one another and their double layers begin to overlap, assuming the Poisson distribution of the space charge is valid, the Poisson-Boltzmann equation may be used as the fundamental relationship to obtain potential as a function of distance of separation between the Stern planes of the particles. Since, physically, the potential distribution is established by ionic species in the double layers of each particle, the potential at one-half the distance of separation ($\psi_{d/2}$) must be equal for both double layers [7].

Thus, the assumptions made to describe the effect of overlapping double layers on potential as a function of distance are:

1. the Poisson-Boltzmann equation is applicable. This means for the Poisson-Boltzmann as given in equation 1, the particles are considered to be semi-infinite flat plates;
2. the distribution of ions between two approaching particles is continuous, and;

3.   the closest the particles may approach one another is their Stern planes.

Implicit to assumption (1) is the premise that the ions may be considered to be point charges which contribute no volume to the solution. However, as the particles approach one another, this assumption may not remain strictly valid [9, 10]. At a finite, but small, distance of separation, probably on the order of 5 Å [7], overlap of ions will occur and the model treating the counter ions as point charges begins to break down. The implication is that at higher concentrations in the bulk solution, more crowding will occur and hence assumption (1) will no longer hold. Fortunately, other deviations from ideality beside crowding appear to negate or reduce this effect. Levine and Bell [10], taking into account ion volume effects, dependence of the dielectric constants on the field strength and electrolyte concentration, the polarization of the ions by the electric field of the double layer, the effect of ionic atmosphere, and the cavity potential which is due to the hydration of counter ions, showed that these effects apparently almost cancel one another, at least for (1-1) symmetrical electrolytes up to concentrations of 0.1 moles/liter. In a much earlier discussion of the volume effects, Grahame [9] theoretically discussed the effect of a finite volume for the counter ions in solutions up to 1 mole/liter. In this less sophisticated approach, it was shown that crowding would not be a problem. For want of a better or more consistently accepted mathematical model, the Poisson-Boltzmann equation will be utilized to describe the effect of overlapping double layers.

When two identical double layers overlap, the potential as a function of distance is modified in the following manner. The potential still approaches zero as distance from the Stern plane when the distance of separation between two particles is infinitely large (or at least nearly so). However, as the particles begin to approach one another with the consequent overlap of their Gouy-Chapman layers, the potential at the midpoint between the two particles takes on an ever-increasing finite value ($\psi_{d/2}$). From the assumption of a continuous charge distribution comes the requirement that at $\psi_{\delta/2}$, $d\psi(x) = 0$. These relationships are summarized in Figure 1. This figure gives the Stern potentials of the overlapping particles at -d/2 and +d/2 where d is the distance of separation and gives $\psi_{d/2}$ at x = 0.

This choice of values on the abscissa allows the first integration of the Poisson-Boltzmann equation to proceed in a straightforward manner [11]. With the boundary conditions that $\psi(x) = \psi_{d/2}$ and $d\psi(x)/dx = 0$ when x = 0, the first integration gives

    Characterization Techniques for the Solid-Solution Interface

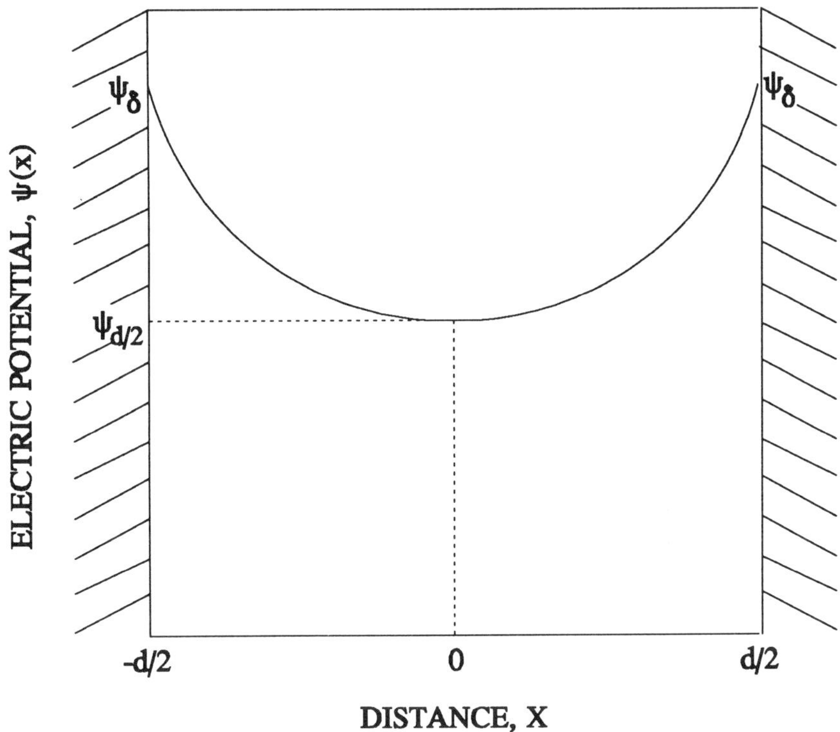

Figure 1 Electric potential as a function of distance, x, for overlapping Gouy-Chapman layers from ψ_δ where $x=\pm d/2$ to $\psi_{d/2}$ where $x=0$ and $d\psi(x)/dx=0$.

$$\frac{d\psi(x)}{dx} = \pm\sqrt{\frac{8\pi kT}{DD_o}}\sqrt{\sum n_i[\exp(-\frac{z_i e\psi(x)}{kT})-\exp(-\frac{z_i e\psi_{d/2}}{kT})]} \quad (4)$$

where $\psi_{d/2}$ is the potential at one-half the distance of separation in volts, and all other symbols are consistent with those given in Equation 1. Equation 4 for overlapping double layers reduces to Equation 2 when $\psi_{d/2} = 0$. Thus Equation 4 may be used as a general expression where $\psi_{d/2}$ varies from zero (e.g., no double layer interaction) to ψ_δ (e.g., complete overlap).

The charge of the Gouy-Chapman layer for overlapping double layers is related to Equation 4 in the same manner as Equation 3 is related to Equation 2. It is given by

$$\sigma_{GC} = \pm\sqrt{\frac{DD_o kT}{2\pi}}\sqrt{\sum n_i[\exp(-\frac{z_i e\psi_\delta}{kT})-\exp(-\frac{z_i e\psi_{d/2}}{kT})]} \quad (5)$$

In Equation 5, as $\psi_{d/2}$ increases (e.g., as overlap of the Gouy-Chapman layers increases), the Gouy-Chapman component of the surface charge decreases. This reduction in σ_{GC} with increasing overlap leads to the repulsive energies generated during increasing interaction of the Gouy-Chapman layers of two particles.

Only overlap of the Gouy-Chapman layers is considered in the analysis of repulsive energies. As a consequence, only that part of the surface charge compensated for by the Gouy-Chapman layer need be considered. In actuality, the surface charge is compensated by the charge in both the Stern and Gouy-Chapman layers. To calculate repulsive energies, it is assumed that Stern potential is constant during the course of the interaction [7]. Only knowledge of the Gouy-Chapman contribution to the surface charge is essential in the analysis.

Before a discussion of the calculation of repulsive energies arising from the overlap of Gouy-Chapman layers is given, it is necessary to present the second integration of the Poisson-Boltzmann equation for overlapping double layers to obtain potential as a function of distance of separation. Unfortunately, Equation 4 does not lend itself to an analytical solution. Therefore, with the boundary conditions that $\psi(x) = \psi_\delta$ when $x = d/2$ and $\psi(x) = \psi_{d/2}$ when $x = 0$ requires the

     Characterization Techniques for the Solid-Solution Interface

following integral equation to be numerically evaluated:

$$d = \int_{\psi_{d/2}}^{\psi_s} \frac{2d\psi}{\sqrt{\dfrac{8\pi kT}{DD_o}} \sqrt{\displaystyle\sum_{i=1}^{\infty} n_i[\exp(-\frac{z_i e\psi(x)}{kT}) - \exp(-\frac{z_i e\psi_{d/2}}{kT})]}} \quad (6)$$

where d is distance of separation between two particles in centimeters for given values of ψ_s and $\psi_{d/2}$. In the limiting cases of $\psi_{d/2} = \psi_s$ and $\psi_{d/2} = 0$, the distance of separations are equal to zero and approaching infinity, respectively. The integrand in Equation 6 is general requiring no assumption as to the number or valence of ionic species.

In Equation 6, when $\psi(x)$ approaches $\psi_{d/2}$, the integrand approaches infinity. Therefore, the relationship in equation 6 is reduced to the sum of two integrals such that

$$d = \int_{\psi_q}^{\psi_s} \frac{2d\psi}{\sqrt{\dfrac{8\pi kT}{DD_o}} \sqrt{\displaystyle\sum_{i=1}^{\infty} n_i[\exp(-\frac{z_i e\psi(x)}{kT}) - \exp(-\frac{z_i e\psi_{d/2}}{kT})]}}$$

$$+ \int_{\psi_{d/2}}^{\psi_q} \frac{2d\psi}{\sqrt{\dfrac{8\pi kT}{DD_o}} \sqrt{\displaystyle\sum_{i=1}^{\infty} n_i[\exp(-\frac{z_i e\psi(x)}{kT}) - \exp(-\frac{z_i e\psi_{d/2}}{kT})]}}$$

$$= I_1 + I_2, \quad (7)$$

where the second integral on the right hand side of Equation 7 (I_2), assuming the difference between ψ_δ and $\psi_{d/2}$ is small, may be approximated by [11]

$$I_2 = \frac{2(\psi_\delta - \psi_{d/2})}{\sqrt{\dfrac{8\pi kT}{DD_o}}\sqrt{\displaystyle\sum_{i=1}^{\infty} n_i\left[\exp\left(-\frac{z_i e \psi_q}{kT}\right) - \exp\left(-\frac{z_i e \psi_{d/2}}{kT}\right)\right]}}.$$ (8)

It can be shown that as $\lim(\psi_\delta-\psi_{d/2})\rightarrow 0$, I_2 converges to zero. Thus, a simple trapezoidal rule may be used to evaluate I_2. Utilizing this approximation, $\psi_{d/2}$ as a function of distance of separation was calculated with Simpson's rule [11] for I_1, and Equation 8 for I_2. The results are summarized in Figure 2. The Stern potential was taken to be 100 mV and the ionic concentration was taken to be 9.254 x 10^{-16} M with valences equal to +1 and -1 to give $\kappa^2 = (8\pi e^2 n)/(DD_o kT)$ = 1 cm^{-1}, where κ is the reciprocal thickness of the double layer and T = 298.16°K with D = 78.55. Thus, κd yields distance of separation in dimensionless units. The results of the numerical integration techniques are indicated by the points drawn. The solid line is obtained using the simplified form of the integral assuming a symmetrical electrolyte [12]. There is excellent agreement between the solutions of the generalized integral and the simplified integral. However, at the limits of $\psi_{d/2}$, there are deviations. Fortunately, these relatively small deviations occur where the attractive energies due to van der Waals interactions dominate the total energy of interaction (e.g., large $\psi_{d/2}$ and small distances of separation), or where the repulsive energies become negligible (e.g. small $\psi_{d/2}$ and large distances of separation).

Repulsive Energies Arising from Double Layer Interaction

When identical double layers begin to overlap, repulsive energy develops [6,7]. However, an individual particle and its associated double layer has no net charge. Therefore, the repulsive energy of two interacting particles cannot be calculated by Coulomb's law. Assuming that thermodynamic equilibrium is maintained during the increasing overlap of the double layers and that the interface consists of a surface charging reaction accompanied by Stern and Gouy-Chapman layers gives a valid assessment of the double layer, then an analysis may be performed to describe the repulsive energy as a function of distance of

   Characterization Techniques for the Solid-Solution Interface

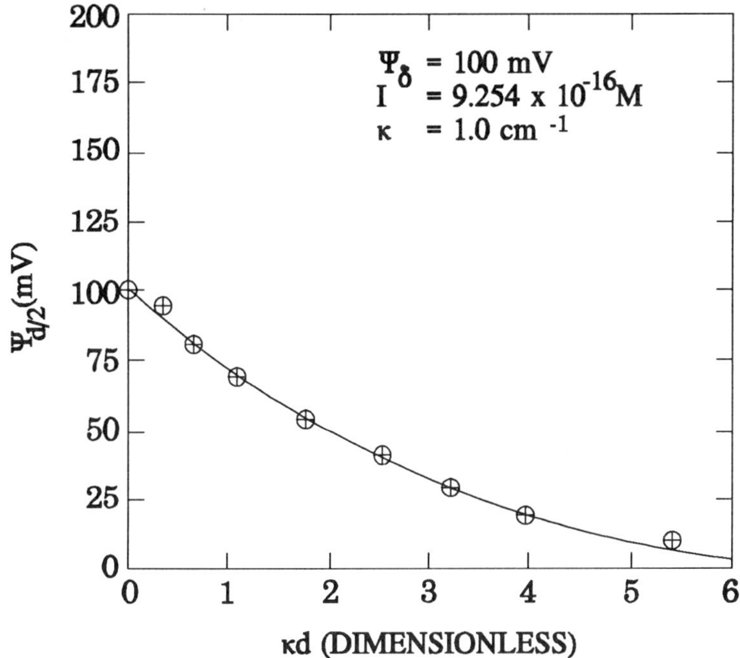

Figure 2 Potential at one-half separation distance as a function of κd. Solid line is obtained from assumption of symmetical electrolytes. Points were obtained from calculation performed by DLVO for general equations. Variables held constant are indicated.

separation for 1 cm^2 of area of two semi-infinite flat plates. While analysis is given paralleling the theoretical procedure given by Verwey and Overbeek [7], all equations developed are general, such that no assumptions are made as to electrolyte valence, concentration, or number of ionic species present.

The general procedure is to obtain the repulsive energy ($V_{R\text{-pl}}$ (d)) at a given distance of separation by taking the free energy difference between the single disturbed double layer and the undisturbed double layer and multiplying by two, since we are considering two interacting double layers. This gives for a given distance of separation and Stern potential the repulsive energy due to the double layer overlap of semi-infinite plates,

$$V_{R\text{-pl}}(d) = 2(F(d) - F(\infty)),\qquad(9)$$

where $V_{R\text{-pl}}(d)$ is in ergs cm^{-2}.

The next step is to incorporate a relationship developed by Deryagin [13] between the repulsive energies as functions of distances of separation for 1 cm^2 of semi-infinite parallel plates and the repulsive energy developed for large spherical particles ($V_{R\text{-sp}}(H_o)$ given by

$$V_{R\text{-sp}}(H_o) = a\pi \int_{H_o}^{\infty} V_{R\text{-pl}}(d) \; dd,\qquad(10)$$

where $V_{R\text{-sp}}(H_o)$ is in ergs, for two spherical particles of radius, a, in centimeters, and H_o is the distance of closest approach for the spherical particles. This equation assumes that the radius of the particles is large with respect to the Gouy-Chapman layer. Implicit in this assumption is that the lines of force remain essentially parallel to the axis of symmetry drawn between the centers of each particle. Verwey and Overbeek [7] showed that these assumptions should be reasonably valid so long as $\kappa a > 2.5$, where κ is the reciprocal thickness of the double layer. Since we are interested in analyzing coagulation behavior of particles in relatively high ionic strength solutions, Equation 10 is usually applicable. Thus, an evaluation of the repulsive energy developed by the interaction of the double

     Characterization Techniques for the Solid-Solution Interface

layers of 1 cm^2 area of parallel semi-infinite plates will enable us to determine the total repulsive energy developed between the interaction of the double layers of relatively large spherical particles.

Verwey and Overbeek developed a method to assess the free energy of interacting double layers based upon concepts of solution thermodynamics established by Debye and Huckel for strong electrolytes [7]. This procedure involves beginning with a double layer in thermodynamic equilibrium which may or may not be interacting with another double layer, but which extends from a distance of zero at the surface to d/2, where $\psi_{d/2}$ is found. The adsorbed layer of ions in the double layer is incrementally removed from the double layer with a corresponding incremental removal of the charge of the ions, while maintaining thermodynamic equilibrium by keeping the surface potential constant. The net result after the removal and discharging process is that a particle with no double layer is obtained. Thus, the work required to remove the ions and discharge them is the free energy of the double layer.

Starting from the equation given by Verwey and Overbeek for the situation where a layer of adsorbed ions is present and we wish to remove from the surface as well as remove their charge, the total free energy created by this "discharging" process may be calculated. This is given by

$$F(d) = \int_0^1 \frac{d\lambda}{\lambda} \int_{x=0}^{x=d/2} \rho'(x)\psi'(x)\ dx, \tag{11}$$

where $d\lambda$ is the increment by which the fraction of the charge of anion has been reduced toward neutrality, $\rho'(x)$ is the space charge at some perpendicular distance, x, from the surface after a certain fraction of discharge, λ, has occurred and $\psi'(x)$ is the potential at some perpendicular distance, x, during the discharging procedure. The parentheses, (x), will be dropped from subsequent equations, but it should be kept in mind that ρ' and ψ' are dependent upon the distance from the Stern plane.

With the equilibrium condition that $\lambda e\psi_\delta + \lambda\Delta\mu = 0$, the chemical potential must be reduced at the same rate as the ionic charges. Thus, for each degree of discharging, λ, we have $-\lambda e\psi_\delta = \lambda\Delta\mu$, so that ψ_δ remains unchanged. For each step of electric work gained, $e\psi_\delta d\lambda$, a corresponding amount of chemical work is performed, $\Delta\mu d\lambda$, in decreasing the charge of a surface ion by the amount $ed\lambda$. The electric and chemical work thus compensate one another. An important result of this is that the energy of the double layer may be calculated if only the electric work is considered. Furthermore, those ions which contribute to the double layer need not be considered, because the net effect of discharging all ions in the bulk solution or the counter ions of the double layer is at most a secondary effect.

Verwey and Overbeek solved Equation 11, but simplified the equations by assuming symmetrical electrolytes were present. This simplification allowed them to solve the corresponding integrals to a point, where without the aid of a digital computer, numerical integration of the result could be performed. However, we wanted to utilize the basic principles of fine particle physical chemistry and mathematics set forth by Verwey and Overbeek, but solve for the general case of any number of ions of any valence present in the bulk solution. For those integrals which prove to be intractable analytically, numerical integration with a digital computer is readily available. Thus, virtually all of the mathematical analysis and steps are directly related to the work of Verwey and Overbeek, except that the more general exponential functions given in Equations 3 and 4 are utilized instead of the hyperbolic function obtained from the simplification arising from the assumption of symmetrical electrolytes. Integration of Equation 11 yields

$$F(d) = -[\frac{kTd}{2}][\sum_{i=1}^{\infty} n_i(\exp(-\frac{z_i e \psi_{d/2}}{kT}) - 1)]$$

$$- \int_{\psi_{d/2}}^{\psi_s} \sqrt{\frac{DD_o kT}{2\pi}} \sqrt{\sum_{i=1}^{\infty} n_i[\exp(-\frac{z_i e \psi(x)}{kT}) - \exp(-\frac{z_i e \psi_{d/2}}{kT})]} \, d\psi. \quad (12)$$

Substitution of d calculated using Equation 6 into Equation 12 will yield F(d). Use of Equations 9 and 10 gives repulsive potentials for interacting flat plates and spheres, respectively, which when combined to the appropriate van der Waals

     Characterization Techniques for the Solid-Solution Interface

interactions yields the interaction energy curves. A computer program, DLVO[**], has been written that performs such calculations.

VERIFICATION OF CALCULATIONS PERFORMED BY DLVO

Total interaction energy curves calculated by DLVO and a conventional method [7] are essentially the same. Using a temperature of 298.16°K, a dielectric constant equal to 78.55, a Lifshitz-van der Waals constant 1×10^{-12} ergs, a particle spherical radius equal to 0.1μm, and a Stern potential equal to 25.6 mV, the effect of the ionic strength on total interaction energy as a function of a dimensionless distance parameter, s, is obtained. The parameter, s, is the center to center particle distance divided by the radius of the particles where $s = 2 + H_o/a$.

Total interaction energy curves calculated using DLVO for different ionic strengths are given in Figure 3. Comparison of Figure 3 with Figure 40, page 161, in The Theory of the Stability of Lyophobic Colloids [7] indicates excellent agreement between the calculations performed by DLVO and the calculations obtained by the numerical analysis of the simplified equations. Only for the cases where $\kappa a < 2.5$ is there poor agreement. This is for $\kappa = 1 \times 10^5$ such that $\kappa a = 1.0$. However, even this deviation is not substantial since the maximum total interaction energies from Figure 40 and Figure 3 are within 1-2 kT of one another. The distance at the maximum is shifted, however. The curves generated for $\kappa = 1 \times 10^5$ cm^{-1}, and $\kappa = 0$ by Verwey and Overbeek utilize an approximate equation for V_{R-sp} (d) in the case where $\kappa a < 2.5$.

Total interaction energy curves calculated using DLVO as a function of Stern potential are given in Figure 4. Debye thickness is 1×10^6 cm^{-1} (e.g., n = 9.254×10^{-4}M). Other variables are the same as given above. Comparison of Figure 4 with Figure 41, page 162, in The Theory of the Stability of Lyophobic Colloids [7] gives, once again, excellent agreement for total interaction energy curves calculated using the two methods.

CONCLUSIONS

Starting from the generalized Poisson-Boltzmann equation, the relationships which make no assumptions concerning the valence or number of types of ions present in the bulk solution have been presented for:

1.   the Gouy-Chapman contribution to the surface charge as a function

---

[**]DLVO is currently available with an operation manual through J.H. Adair, 207 MAE building, University of Florida, Gainesville, FL 32611

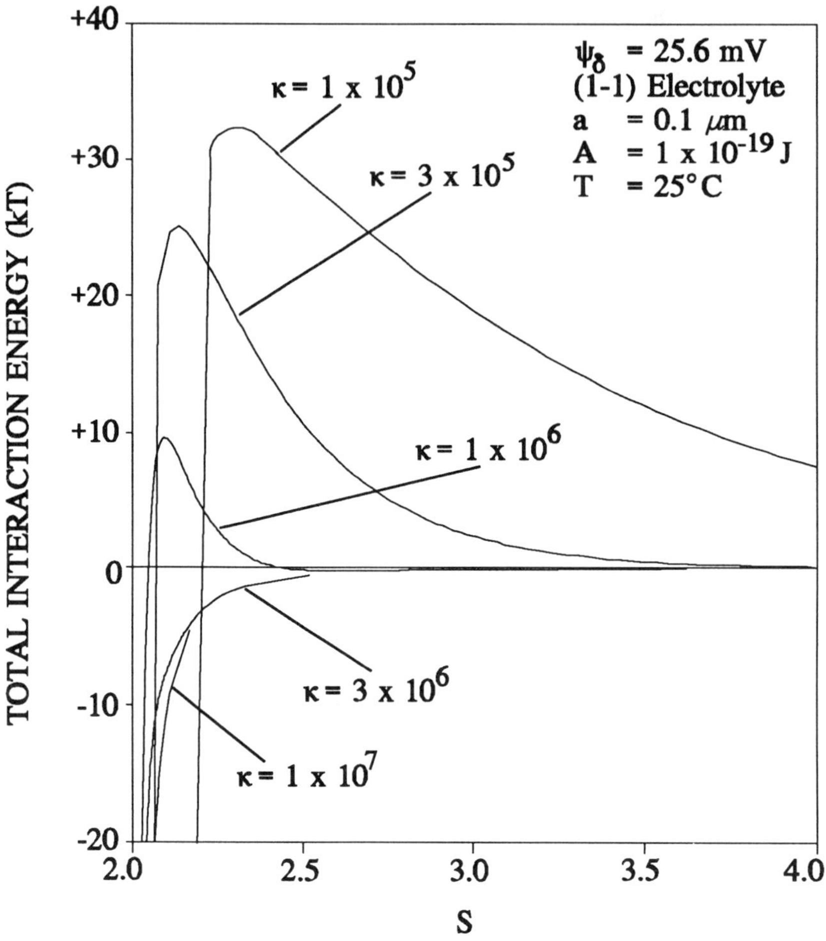

Figure 3 Total interaction energy curves as a function of ionic strength. Variables held constant are indicated. The spherical model was used for calculations.

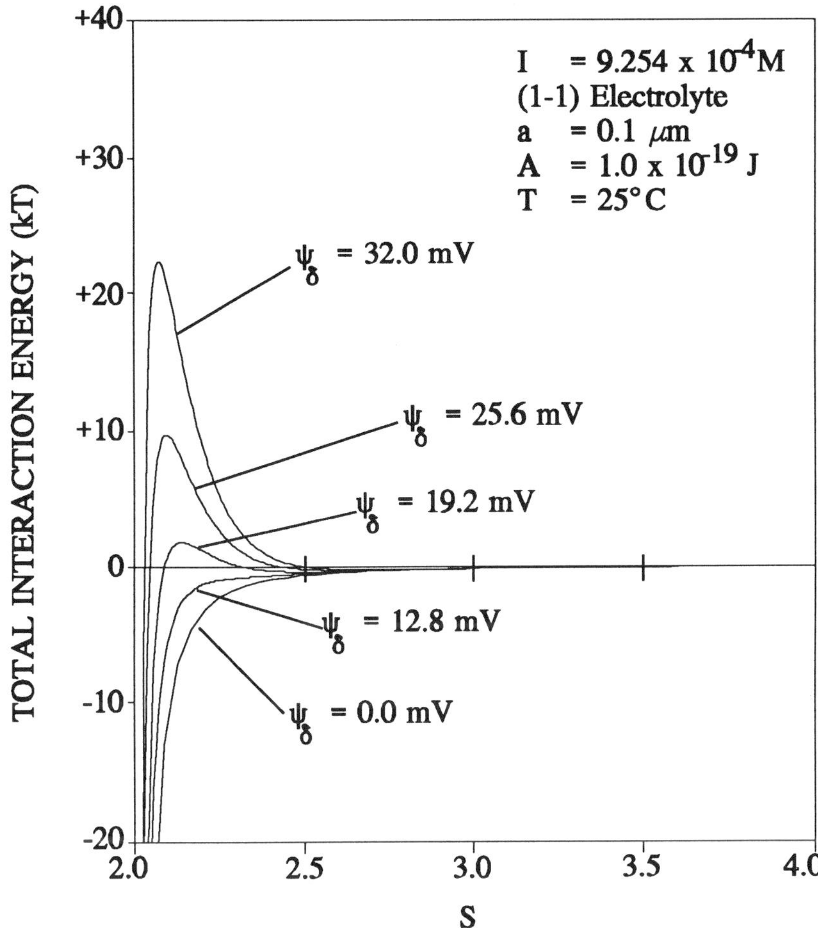

Figure 4 Total interaction energy curves as a function of the Stern potential. Variables held constant are indicated. The spherical model was used for calculations.

of $\psi_{d/2}$ for a given Stern potential;

2. distance of separation of two semi-infinite flat plates as a function of potential at one-half the distance of separation, $\psi_{d/2}$, for a given Stern potential; and

3. the free energy of the single double layer for the overlapping or the undisturbed situation as a function of distance of separation for two semi-infinite plates.

Furthermore, the general procedure utilized to calculate the total electrostatic energy contribution to the total interaction energy has been summarized.

Calculations made using DLVO and conventional methods were in excellent agreement. However, the more versatile general equations in DLVO may be utilized for more complex solutions than those found in symmetrical electrolyte solutions.

ACKNOWLEDGEMENT

The authors gratefully acknowledge the partial support of this work by National Institutes of Health POG 5P01 DK20586-17.

REFERENCES

1. Overbeek, J.Th.G., J. Colloid Interface Sci. 58, 408 (1977).
2. Matijevic, E., J. Colloid Interface Sci. 43, 217 (1973).
3. Ottewill, R.H., "Colloid Science: Specialist Periodical Reports," Chemical Society, London, 173 (1975)
4. Visser, J., Adv. Colloid Interface Sci. 2, 396 (1969).
5. Hunter, R.J. "Foundations of Colloid Science," Oxford Science Publications, Oxford, 1986.
6. Deryagin, B.V. and Landau, L., Acta Physicochem. URSS 14, 633 (1941).
7. Verwey, E.J.W. and Overbeek, J.Th.G., "Theory of the Stability of Lyophobic Colloids," Elsevier Publishing Co., Inc., Amsterdam, 1948.
8. Loeb, A.L., Overbeek, J.TH.G. , and Wiersema, P.H., "The Electrical Double Layer Around a Spherical Colloid Particle," M.I.T. Press, Cambridge, Mass, 1961.
9. Grahame, D.C., Chem. Revs. 41, 441 (1947).
10. Levine, S. and Bell, G.M., Discussion Faraday Soc. 42, 69 (1966).
11. Burden, R.L. and Faires, J.D., "Numerical Analysis," PWS-KENT Publishing Company, Boston, 1989.
12. Sonntag, H. and Strenge, K., "Coagulation and Stability of Disperse Systems," Halsted Press, New York, 1972.
13. Deryagin, B., Kolloid-Z. 69, 155 (1934).

Electrokinetic Methods

MODEL HYDROUS OXIDE COLLOIDS FOR PARTICLE ELECTROPHORESIS

W.A. Zeltner[1], J.-F. Wang[2], O.O. Omatete[3], M.A. Janney[3], M.I. Tejedor-Tejedor[1], M.A. Anderson[1], R.E. Riman[2], D.J. Shanefield[2] and J.H. Adair[4]

[1] University of Wisconsin, Water Chemistry Program, 660 N Park St., Madison, WI 53706
[2] Rutgers University, Department of Ceramics, P.O. Box 909, Piscataway, NJ 08855
[3] Oak Ridge National Laboratory, P.O. Box 2008, Oak Ridge, TN 37831
[4] University of Florida, 323 MAE Building, Gainesville, FL 32611

ABSTRACT

Although several commercial instruments are available for measuring the electrophoretic mobility of particles, standard materials for calibrating these instruments are lacking. While individual batches of latex particles can serve as model colloids, batch-to-batch reproducibility of latexes is difficult to attain. Latexes also do not exhibit an isoelectric point, an important characteristic of aqueous suspensions of ceramic materials. Since hydrous oxide colloids display isoelectric points, they may serve as calibration standards for measuring electrophoretic mobilities and determining isoelectric points. Aqueous suspensions of phosphated goethite (α-FeOOH) have shown evidence of providing an "absolute" mobility measurement. The suitability of this material as a standard for particle electrophoresis was further evaluated by performing round-robin measurements of pH and mobility among three different laboratories using different instruments. Suitable protocols for obtaining the needed data were developed to minimize variations in the measurements.

INTRODUCTION

In particle electrophoresis, one measures the velocity of a particle in an applied electric field. The resulting parameter (velocity/electric field) is called the mobility of the particle and is very useful for characterizing the net electric charge and the electrostatic potential of suspended particles with respect to the suspending phase. For particles whose charges vary with the pH of the suspension, particle electrophoresis provides a relatively straightforward means of determining the isoelectric point (IEP) of the system, the pH at which the particles carry no net charge. Both the particle mobility and the IEP are useful for studying phenomena such as particle aggregation, the stability of suspensions, the adsorption of ions on particles, and the development of electrical double layers around particles.

A major drawback of the electrophoresis technique is that particle mobility is measured at a surface of shear or slipping plane, a surface surrounding the particle within which the suspending fluid is stationary with respect to the particle. Since the position of the slipping plane is poorly defined, it is difficult to incorporate data from particle electrophoresis into electrical double layer models. In spite of this limitation, the information about charge development on particles that can be obtained from particle electrophoresis studies is very useful and easily obtained. Thus, this technique has become an extremely important tool for researchers working in such industries as ceramics and pulp and paper processing. A particularly useful treatment of particle electrophoresis and the associated double layer phenomena can be found in Hunter (1).

Several companies market instruments for performing particle electrophoresis (e.g., PenKem, Inc., Malvern Instruments, Inc., Coulter Corp.). While most of these instruments are based on photon correlation measurements, the PenKem unit utilizes a separate optical technique to measure particle mobility. Although these units have been in use for many years, we are not aware of any published studies that compare measurements obtained on different instruments for the same material. The only assurance researchers have that measurements obtained on one instrument could be reproduced on another instrument is that obvious measurement errors have not been reported. For example, one could take an aqueous suspension of pure silica at pH 6 and, at the appropriate particle concentration, confidently expect to obtain a negative mobility for those particles on any commercially available instrument. If the measured value were positive or just slightly negative, one would suspect either sample contamination or instrument misalignment or malfunction. However, even if a relatively large negative mobility were measured for this sample, one would not know how much variation to expect in that measured value if the same sample was analyzed by another researcher using another instrument.

A related issue is that of standards for use in calibrating these instruments. Ideally, one would like to have a material with a known mobility <u>and known tolerances</u> that could be used to check the performance of particle electrophoresis units. In practice, however, the development of such standards is a complex matter involving the choice of a suitable candidate material and its evaluation in inter-laboratory round robin studies.

Another issue arises in the ceramics community, which uses IEP measurements as a particle characterization method. In order to obtain an IEP for an aqueous suspension of particles, the mobilities of those particles must be measured at pH values that bracket the IEP. Once again, one is faced with the question of determining if variations in the performance of commercial instruments could affect the IEP value. One also would like to know what tolerance to expect in such measurements across different instruments. This issue is related to that of developing standard materials for particle electrophoresis but is further complicated by the necessity of evaluating both pH and mobility measurements.

          Characterization Techniques for the Solid-Solution Interface

In a joint project, researchers from Oak Ridge National Laboratory, Rutgers University, the University of Wisconsin - Madison and the University of Florida at Gainesville agreed to study the surface reactions of silicon nitride in suspensions and also to investigate the issues raised above. This phase of the study had three major objectives:

1. to select an appropriate material as a possible standard for performing particle electrophoresis measurements;

2. to evaluate the performance of the chosen material in a round robin comparison of its behavior among the four separate laboratories;

3. to apply the protocols developed in completing the second objective to a round robin comparison of the particle electrophoresis of aqueous suspensions of silicon nitride.

In this paper, we report on the criteria used to select a material for evaluation as a possible standard for particle electrophoresis, the experimental protocols that were developed to help assure the best reproducibility in performing the required measurements, and preliminary results of the round robin evaluation of the selected material. A detailed statistical analysis of these results will be presented separately.

SELECTION OF MATERIALS

Ideal Requirements for a Standard Material

An ideal standard for measuring particle mobility should possess the following properties (not all-inclusive and not listed in order of importance).

1. Cheap
2. Spherical
3. Monodisperse
4. Commercially available
5. Insoluble under measurement conditions
6. Samples easily prepared
7. Reproducible measurements within one sample
8. Reproducible properties among different samples of the same material (batch-to-batch reproducibility)
9. Narrow mobility distribution (to ensure confidence in the reproducibility of the measurement)
10. Temporal stability (long shelf life with no aging)
11. Provides an "absolute" mobility (the value correlates with data obtained by other techniques)
12. Possesses an IEP
13. IEP adjustable to desired pH range
14. Mobility adjustable over a wide range of positive and negative values
15. Scatters He-Ne laser light well
16. Forms relatively stable suspensions (particles do not settle out or settle out relatively slowly during the measurement)

17. Sample behaves in an expected manner (at a given pH, mobility decreases as the ionic strength of the suspension increases)

Clearly it is unlikely that any one material could satisfy all of these requirements, especially since the relative importance of each of these properties will vary depending on the particular application. For instance, if one wants to calibrate an instrument over a narrow range of mobility or pH values, then possession of an IEP is unnecessary. If enough demand exists for a mobility standard and an appropriate material is selected, then the National Institute of Standards and Technology (NIST) could consider preparing a standard reference material. In this case, NIST might want to synthesize the standard themselves, and commercial availability of the material would be irrelevant.

Materials Considered As Possible Standards

Several materials were evaluated as candidates for a round robin study of mobility measurements, including latexes, red blood cells, and two particular hydrous oxides - silica and goethite (α-FeOOH). Emulsions were not considered only because we had no familiarity with their properties and so did not feel competent to evaluate such materials. Thus, an emulsion could be a viable standard for calibrating particle electrophoresis units.

Two of the listed criteria were major factors in choosing a material for further evaluation. First, possession of an IEP was important because the protocols developed during the evaluation of a candidate standard material would be applied to the later determination of the agreement among laboratories in measuring the IEP of aqueous suspensions of silicon nitride. Second, we preferred to use a material that had shown evidence of providing an "absolute" measurement of mobility. Given the goals of this study, several criteria were not considered important at this stage. These criteria included commercial availability (if enough material was readily available), batch-to-batch reproducibility (the behavior of one batch of material was evaluated during the round robin), long shelf life (beyond the one month period needed to conduct the round robin), and adherence to expected behavior. Given these constraints, the evaluation of each candidate material will be discussed next.

Red Blood Cells

Red blood cells have advantages as standards for measuring mobility. They can be obtained very cheaply (by pricking a finger), display narrow mobility distributions, may give reproducible results, and provide a range of mobilities (either positive or negative) depending on the buffer used to prepare the suspension. However, these advantages are offset by numerous disadvantages. Because red blood cells are only viable over a small pH range, they can only be used to calibrate particle electrophoresis units over that small pH range. There are severe health concerns in obtaining samples by pricking a finger. Sample degradation is rapid, so samples must be prepared fresh before each use. Red blood cells are relatively large and may settle out before readings can be obtained on

    Characterization Techniques for the Solid-Solution Interface

some instruments. We know of no "absolute" measurements of mobility for red blood cells. Thus, red blood cells were not seriously considered as a possible mobility standard.

Latexes

Latexes have several advantages over red blood cells as possible mobility standards. Latexes are spherical, commercially available and easily prepared as stable suspensions. Individual samples can cover a wide range of pH and mobility values. If care is taken in preparing and storing samples, they should have a relatively long shelf life. However, there are some concerns that particle aggregation can occur in latex suspensions during storage which can then affect mobility measurements. Although not experts in this area, we assume that sonication of samples and treatment with small amounts of biocides would limit such problems. Latexes often deviate from expected behavior with changes in ionic strength for reasons that are not well understood. (The effect of ionic strength on the mobility of latexes is an active research area at present.) Short-term measurements on aqueous suspensions of latexes show that such materials can provide narrow mobility distributions and good reproducibility of measurements for samples prepared from a given batch of latex (2).

While latexes can be considered for use as mobility standards, they have disadvantages. (1) Most latexes contain only one type of reactive surface group that can be either charged or uncharged. Thus, these materials do not display an IEP. (Some latexes contain two different surface groups, one positive and one negative, so such materials might be suitable for IEP measurements.) (2) Based on comments from several individuals, batch-to-batch reproducibility of latexes appears to be a problem. Suspensions prepared from different batches of supposedly identical latex samples can display different mobilities. (3) We are not aware of any "absolute" mobility measurements obtained with a latex.

We feel that suspensions of latex particles prepared from individual batches could be utilized as standards for particle electrophoresis when one desires to calibrate an instrument under one or two distinct sets of conditions. However, the question of batch-to-batch reproducibility of the mobilities of different batches of latex particles with otherwise identical properties must be addressed. Latex suspensions would not be very suitable for use as standards to evaluate a procedure for determining an IEP. Since IEP measurements were needed for our study, latexes were not included in the round robin.

Hydrous Oxides - General Considerations

Because most hydrous oxides display clear isoelectric points, they are obvious candidates for this study. However, the large number of available hydrous oxides makes it very difficult to select a suitable material. We opted to limit our choices to materials which had been investigated in our laboratories. Two materials were considered for evaluation in the round robin study, silica prepared by the Stöber process and goethite (α-FeOOH).

Although most hydrous oxides possess a clear IEP, they have a major drawback to their use as mobility standards. Generally, the surfaces of freshly prepared hydrous oxides are quite reactive. This surface reactivity leads to long-term aging effects that can be observed when aqueous suspensions are prepared or when their pH is changed. Before a suspension of a freshly prepared hydrous oxide could be used as a mobility standard, its aging behavior would have to be studied carefully. Alternatively, the suspension could be allowed to age (we recommend a minimum of <u>one month</u> of aging) before its properties were determined. Even then, further aging studies would need to be performed.

Hydrous Oxides - Silica

Silica particles prepared by the Stöber process (3) are similar to latex particles in their suitability as mobility standards. Stöber silica particles are nearly spherical and monodisperse. Stable suspensions of these particles are easily prepared. However, the solubility of silica increases dramatically above pH 8.5, which would probably preclude the use of these particles under very basic conditions. More seriously, silica does not display an obvious IEP. Suspensions of pure silica have negative mobilities at pH values above about 3. Below pH 3, the mobilities are small, and silica suspensions do not display highly positive mobilities. One should obtain positive mobilities for silica, and thus a measurable IEP, by specifically adsorbing a cation on the silica surface, but we do not know how much of a shift in mobility to expect. We also do not know how severe a problem aging is for this material. Finally, we are not aware of any "absolute" measurements of mobility with Stöber silica. Because of a lack of experience in characterizing suspensions of Stöber silica and concerns about obtaining a clear IEP from such suspensions, silica was not considered as a mobility standard.

Hydrous Oxides - Goethite

Goethite was suggested as a possible mobility standard by researchers at the University of Wisconsin (UW), who have been characterizing goethite suspensions for several years. Although electron photomicrographs indicate that the goethite particles are shaped like bricks (ca. 60x20x15 nm), the particles have a fairly narrow size distribution and an IEP between 9.5 and 10.0 (4,5). However, addition of appropriate amounts of phosphate to these suspensions can lower the IEP to 4.0 or to any desired intermediate value (6). Reproducible measurements of mobility with narrow distributions were readily obtained in these studies. Since goethite is not very soluble above pH 3 (samples of goethite held at pH 4.0 for two months do not generate measurable amounts of dissolved Fe^{3+} (7)), mobility measurements can be performed over a wide pH range. In addition, there was evidence that goethite can provide an "absolute" mobility measurement that correlates well with data obtained by other techniques (6).

Tejedor-Tejedor and Anderson (6) obtained phosphate adsorption isotherms on goethite. For given phosphate loadings, they used cylindrical internal reflection - Fourier transform

     Characterization Techniques for the Solid-Solution Interface

infrared spectroscopy measurements to determine the structures of the adsorbed phosphate species. From these measurements, they obtained apparent pK values for deprotonation of the surface species. These values were considered apparent because they varied with suspension pH, which suggests that they were affected by the charge on the particles. Electrophoretic mobilities were measured to characterize the particle charge. At the IEP for each level of phosphate adsorption (100, 150 and 190 μmol PO_4/g), no electrostatic forces exist, and the apparent pK for surface deprotonation is equal to the intrinsic pK for this reaction. This value was estimated at 4.6. Mobilities at pH values other than the IEP were converted to zeta potentials using the program of O'Brien and White (8). The effect of electrostatic interactions was assumed to be given by a Boltzmann expression, $2.3e\psi_s/kT$, with ψ_s equal to the calculated zeta potential.

These workers realized that the validity of these assumptions was debatable. However, after incorporating the Boltzmann expression in the analysis, the corrected pK values varied between 4.5 and 4.8 for all the adsorption levels and pH values studied. This agreement between the corrected pK values and the intrinsic pK for the surface reaction suggests that the assumptions made were valid and also suggests that the mobilities that were measured were "true" values independent of the measuring instrument. Thus, phosphated goethite might serve as an "absolute" standard for mobility measurements. Based on this observation and the ease with which the IEP of goethite could be changed by adjusting the phosphate loading, phosphated goethite was selected for further evaluation in this round robin study of particle mobility and IEP measurements.

PROCEDURE FOR PERFORMING THE ROUND ROBIN

Synthesis of Goethite

Goethite was synthesized by the method of Atkinson et al. (9), using an OH:Fe ratio of 2 (7). Atkinson et al. demonstrated that the size of the goethite crystals produced during the alkaline hydrolysis of ferric nitrate solutions depended on the OH/Fe ratio employed for the initial hydrolysis. As this ratio was increased to 2, the particle size progressively decreased. The goethite used in the present study was taken from a large batch of material that had been prepared several years earlier and stored as an aqueous suspension in a plastic container. One complication in this study is that the characterization data presented above was obtained from a different batch of goethite that had been synthesized at the same time by the same method but had been freeze dried before use. Thus, the properties of the freeze dried goethite and the suspended goethite were first compared to verify that the two materials were similar. An advantage of using the suspended goethite is that further hydration and aging of the surface of the goethite was not likely given that the solid had been in contact with water for several years. However, sample aging could occur with freeze-dried goethite because its surface properties could change as it hydrated.

Preliminary Experiments

Preliminary experiments were conducted to learn more about sample aging and the reproducibility of measurements in aqueous suspensions of phosphated goethite. Surface area measurements on both types of goethite agreed very well (80 m^2/g). Phosphated goethite samples were prepared from the goethite stored in suspension following the protocols given in Tejedor-Tejedor and Anderson (6). Mobilities for these samples agreed well (within 0.2×10^{-8} $m^2V^{-1}s^{-1}$) with the mobilities reported by Tejedor-Tejedor and Anderson (6), who had employed the freeze-dried goethite.

In a further test, a preliminary round robin was performed using suspensions of phosphated goethite. The results indicated that relatively good agreement could be obtained for mobility measurements conducted by different groups using different instruments. This study also verified that sample aging was not a problem. Based on these considerations, phosphated goethite was employed in a more detailed round robin study, called the Goethite Electrophoretic Mobility Study (GEMS), although the sample preparation was modified from that used by Tejedor-Tejedor and Anderson (6).

Considerations in the Design of GEMS

Because mobilities of phosphated goethite samples depend on pH, both pH and mobility must be measured, which complicates the study. One reason for performing the preliminary round robin was to obtain data on which to base further decisions about the design of this study. Several considerations affected that design.

1. Although sample aging was not expected, a batch experiment was preferred. Each participant received five separate containers of each sample, which had equilibrated for at least one week, and each container was used only once during the study. The alternative approach would have been to have the participants titrate stock suspensions, making pH and mobility measurements after each addition of titrant. This alternative was rejected because reequilibration of the adsorbed phosphate to the new pH conditions might have taken several hours, which would have complicated performance of the round robin.

2. In the preliminary round-robin, it appeared that more reproducible results were obtained when samples were sonicated before measurements were taken. As a result, a sonication step was included in the measurement protocol for GEMS.

3. Two different instruments were used to measure mobilities, the PenKem System 3000 (at Oak Ridge National Laboratory (ORNL) and UW) and the Malvern Zetasizer II (at Rutgers University (RU)). Each participant followed the same protocol for performing mobility measurements within the limitations of each instrument. The number of spectra accumulated per measurement and the cleaning procedure for the measurement cell were not specified. Each measurement was obtained on a fresh sample, as UW had observed changes in mobility between samples which had sat in the measurement cell for several minutes and samples which had been freshly introduced into the cell (usually no more than 0.3×10^{-8} $m^2V^{-1}sec^{-1}$). This problem was not observed at RU or ORNL.

          Characterization Techniques for the Solid-Solution Interface

4. Participants were encouraged to use an Orion Ross Sure-Flow combination electrode (Model 8172BN) for performing pH measurements. However, this electrode can be difficult to use. The internal electrical connections are fragile, salt buildup in the reference electrode compartment can prevent flushing of the electrode or can crack the outer case when one forces the electrode to flush, and the ground glass joint must be well seated or leakage of the reference electrode filling solution will contaminate the sample. Thus, other electrodes were also used in this study. UW used the Orion electrode, ORNL used a Corning Rugged Body glass combination electrode (Model 476561) and the Orion electrode, and RU used a Metrohm EA107/6.0102.100 pH electrode and a Metrohm EA440/6.0726.100 reference electrode.

5. To minimize problems with CO_2 absorption, all samples were prepared at pH 7 or less. All pH measurements were performed on stirred samples in containers which minimized contact with air (narrow necks and/or tops covered with Parafilm). The use of an inert gas blanket over the sample while taking measurements was not employed. It was feared that loss of CO_2 from the sample, which would be equilibrated with any CO_2 in the sample container, would change the pH. A criterion was also specified for assuming that an equilibrated pH reading had been obtained. This criterion was a compromise between the need to perform the measurements in a reasonable time and the need to wait for true equilibration of the sample and the electrode.

6. During sample preparation at UW, it became clear that the Orion electrode used there performed best when it was conditioned to the samples before being used to take readings. Thus, a conditioning step was included in the protocol for performing pH measurements. A similar conditioning step was also used in the mobility measurements.

7. Possible effects of shipping on the samples were monitored by shipping extra samples at the highest pH from UW to ORNL and then back to UW. Measurements at UW on these samples gave essentially identical results to measurements performed on samples which had not been shipped.

Preparation of Samples

Fifteen different suspensions (five subsamples of each suspension) were prepared and distributed for GEMS. All samples were aqueous suspensions of phosphated goethite at concentrations of 100 μmol KH_2PO_4/g goethite and 100 mg goethite/L suspension (0.0023% by volume). Because of problems in dispensing goethite from the stock suspension, the actual goethite concentration was somewhat less than given above, thus the phosphate loading was somewhat higher than calculated. While the error is difficult to estimate, the actual values should be within 10% of the desired values.

Each suspension was prepared in bulk (ca. 4 L) at one of five nominal pH values (3.5, 4.4, 5.2, 6.1 or 7.0) and one of three ionic strengths (0.001, 0.01 or 0.05 M KNO_3). To prepare the 0.001 M KNO_3 suspension at pH 3.5, the final pH was assumed to be exactly 3.50 due solely to HNO_3, then the additional amount of solid KNO_3 necessary to give a total ionic strength of 10^{-3} M was added to the system. For all other suspensions, the final ionic strength was assumed to be due solely to the added KNO_3. The conductivity of the 10^{-3} M suspension at pH 3.5 was nearly double the conductivities of the 10^{-3} M suspensions at the other pH values due to the high mobility of the proton.

The pH of the bulk suspensions was adjusted by adding either HNO_3 or KOH, allowing the suspension to equilibrate at room temperature for at least 3 hours (usually overnight), pouring a subsample of the suspension into a 30 mL container, and measuring the pH. All pH 3.5 and 4.4 samples equilibrated near the desired pH values, as did the two higher ionic strength samples at pH 5.2. However, pH adjustment of the other samples proved very difficult, probably due to CO_2 absorption effects at the higher pH values. During the round-robin, measured values of pH for the nominal pH 7.0 samples were consistently between pH 6.5 and 6.7.

Once a reasonable pH value had been attained, the bulk suspension was divided among 20 separate pre-cleaned clear HDPE Nalgene containers (I-Chem Research, Catalog #N211-0250), which were then capped tightly. A stream of dry nitrogen was blown over the surface of the pH 7 samples for 30 sec before capping in order to remove most of the CO_2 present in the container.

Experimental Protocols

Due to the number of samples being measured, two days were allotted for performing each of four sets of measurements for GEMS: 29 & 30 Nov., 3 & 4 Dec. (several UW samples were measured on 5 Dec.), 10 & 11 Dec., and 17 & 18 Dec. of 1990. All measurements were conducted according to the following protocols.

SAMPLE PREPARATION
1. Sonicate each sample in its sample bottle with the cap on for 15 min in bath sonicator. Sonicate each sample individually rather than doing 2 or 3 samples together.
2. Let samples stand at least 10 min at room temperature before making pH measurements. (Ignore this step if pH measurements are made using automatic temperature compensation.)

pH METER STANDARDIZATION
1. Standardize using pH 4 and 7 (or near 7) buffers. Change the pH 7 buffer at least weekly.
2. The pH value should be stable to within 0.01 pH unit (0.6 mV) for at least 1 minute.
3. Restandardize the electrode every 4 hours.
4. Condition the electrode to the samples by placing the electrode in a sample for 10 min, then discarding the sample. Do not record the pH value. (Ignore this step if the first sample is at pH 3.5 or 4.4).

          Characterization Techniques for the Solid-Solution Interface

pH MEASUREMENTS

1. Transfer some sample to a container which minimizes contact between the sample and the surrounding air.

2. Set stirring speed roughly between 100 and 150 rpm.

3. Record pH value only after it is stable to within 0.01 pH unit (0.6 mV) for 30 sec. It will probably take several minutes for the samples near neutrality to stabilize.

MOBILITY MEASUREMENTS

1. Condition measurement cell by exposing it to some sample and flushing it with a fresh loading of that sample before making a measurement.

2. For a given sample, record mobility values for 5 fresh loadings of that sample. More measurements may be taken if desired but must be reported separately.

RESULTS AND DISCUSSION

The first two figures are based on mean values of mobility obtained by averaging five separate mobility measurements for a given system. The third figure is based on statistical analysis of the individual mobility values. Because mobilities for this material depend on pH, this study required concurrent measurements of pH and mobility. Thus, the following discussion deals with three separate but related areas: pH measurements, mobility measurements, and isoelectric points (IEP) for the phosphated goethite. The IEP determination incorporates both of the other measurements. Aging of the samples was not observed.

pH Measurements

The bars shown in Figure 1 indicate the range of pH values measured for a given sample (e.g., pH values measured for the sample at nominal pH 7.0 in 0.001 M KNO_3 fell between 6.20 and 6.66, with a possible outlier at 6.88). Clearly, the variation in measured pH values for a given sample depended on the pH and ionic strength of that sample. In poorly buffered systems (pH near neutral and low ionic strength - 0.001 M KNO_3), the difference between the highest and lowest reported pH values was as much as 0.68 pH units. In more highly buffered systems (pH below 5 and high ionic strength - 0.05 M KNO_3), this difference dropped to 0.10 pH units.

Part of this variation may be caused by differences in the behavior of the pH electrodes used by the various groups for this study. The pH measurements in all 0.001 and 0.01 M KNO3 systems by one group were almost always lower than the pH values measured for the same systems by the other groups. This difference does not appear to be caused by errors in measuring the pH because all groups reported similar pH values in all 0.05 M KNO_3 systems. Although our initial statistical studies suggest that these differences are not statistically significant, it would be prudent for workers who use more than one type of pH electrode to compare the performance of their different electrodes on a given sample to verify that all of their electrodes behave similarly.

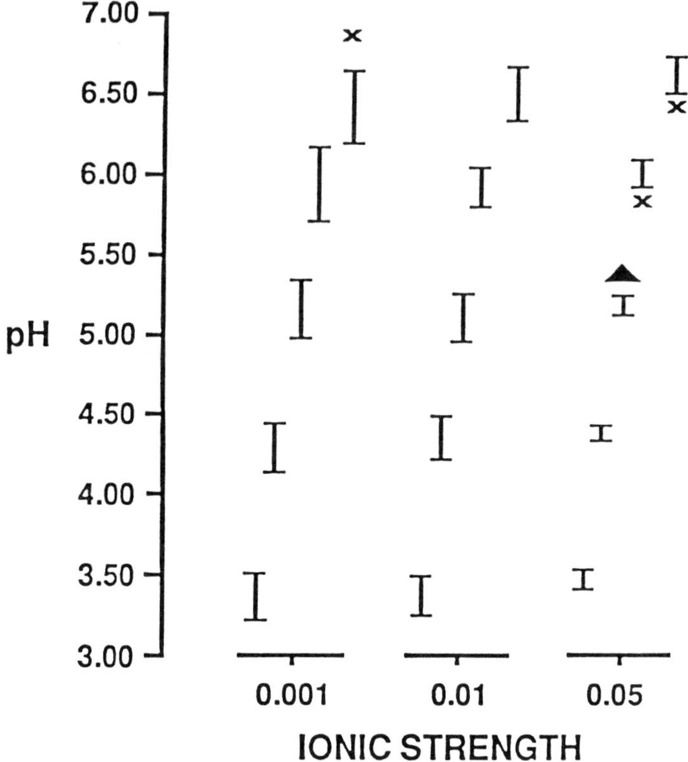

Figure 1. Range of pH values measured for each of 15 samples in GEMS. Each x represents a possible outlier value measured at UW. The solid triangle represents a possible outlier value measured at ORNL. Within a given ionic strength, the range bars are offset from each other to avoid possible overlap.

Mobility Measurements

The bars shown in Figure 2 indicate the range of the mean mobility values measured for a given sample. Although more possible outlier values were obtained for the mobility measurements than for the pH measurements, the variations in the mean mobility values appear to be independent of pH and ionic strength. Differences between the highest and lowest mean mobility values were 0.30 to 0.40 x 10^{-8} $m^2V^{-1}sec^{-1}$ for almost all samples. The large number of possible outliers observed when performing measurements on the 0.001 M ionic strength system suggest that measurements with both types of instruments can be more difficult at low ionic strengths. There were also problems in measuring mobilities with the Malvern unit at the particle concentration used in this study (100 mg

     Characterization Techniques for the Solid-Solution Interface

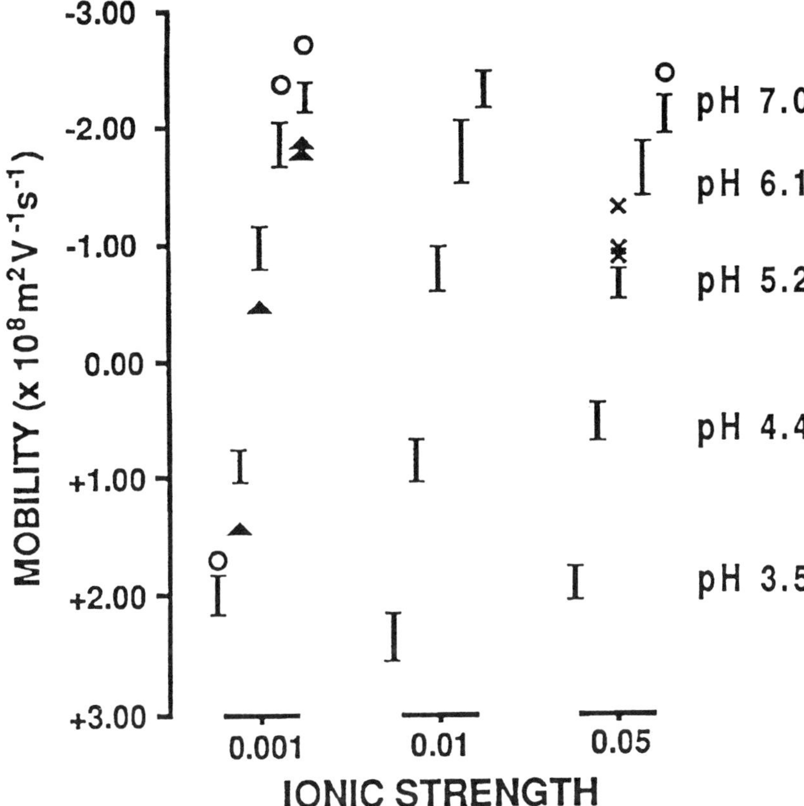

Figure 2. Range of average mobility values measured for each of 15 samples in GEMS. Each x represents a possible outlier value measured at UW, each solid triangle represents a possible outlier value measured at ORNL and each circle represents a possible outlier value measured at RU. Within a given ionic strength, the range bars are offset from each other to avoid possible overlap.

solid/L or 0.0023 vol%). This solids loading had been chosen because it gave good measurements of mobility in the PenKem unit.

In general, electrical double layer models predict that, at a given pH, the electrostatic potential around a suspended particle will increase as the ionic strength of the suspending medium decreases (10). Thus, one would expect the mobility displayed by that particle to also increase with decreasing ionic strength. Inspection of Figure 2 indicates that this phosphated goethite system does not behave in this manner. Instead, in this system the mobilities of the particles are essentially identical in either 0.001 M or 0.01 M KNO_3. The reason for this discrepancy is not known.

IEP Determinations

Since the samples used in this study bracketed the expected IEP, IEP values could be determined by plotting the mean mobility as a function of pH for each ionic strength measured. Given that the three participating groups each performed four sets of measurements at three ionic strengths, 36 IEP values could be obtained by analyzing the data in this manner. All 36 IEP values fall between 4.50 and 5.00 and all but two of the values fall in the range of pH 4.75 ± 0.2. There was no obvious dependence of the IEP on the type of instrument used for the mobility measurements. The small differences in IEP among the three groups may be due to variations in pH measurements. Note that these differences in IEP should be larger if the IEP of the material was closer to pH 7, since we would then expect larger differences in measured pH values among the groups.

IEP values can also be determined from each separate mobility measurement, as shown in Figure 3. Rather than displaying each data point, which would give several black areas in the figure and be uninformative, the data has been represented using variable cellulation. The symbols in this figure represent the number of separate measurements that lie within a given range of pH and mobility. Large symbols with several lines represent many overlapping data points, while individual points are shown by the small symbols. Note that the range of mobilities at a given pH is now larger than shown in Figure 2 because the mobility data has not been averaged. IEP values can be estimated by performing various regressions on this data. If the data from all groups at a given ionic strength is combined, then an overall IEP at each ionic strength can be calculated. This procedure has been employed in Figure 3 using a third order regression. Only a small variation of IEP with ionic strength was observed, between 4.60 and 4.95, for any form of regression.

CONCLUSIONS AND RECOMMENDATIONS

Our results demonstrate that suspensions of phosphated goethite that have been well-aged might serve as a useful standard for determining both mobilities and IEP values. By using this material, we have shown that pH and mobility measurements can be performed reproducibly by workers at different sites <u>in acidic systems</u> if the measurement conditions are carefully controlled. The variation in measured pH values can be as low as 0.10 pH units in highly buffered systems but will be much worse in systems that are poorly buffered (low ionic strength and near neutrality). The variation in measured mobility values is typically $0.3\text{-}0.4\text{x}10^{-8}$ $m^2V^{-1}sec^{-1}$ under most conditions. However, measurements were not performed at the IEP where one often observes greater variability in mobility values. Possible outlier values were observed more often in the low ionic strength systems. Thus, it appears that both the PenKem System 3000 and the Malvern Zetasizer provide identical measurements under acidic conditions. We consider our observed agreement among IEP values to be very good.

　　　　Characterization Techniques for the Solid-Solution Interface

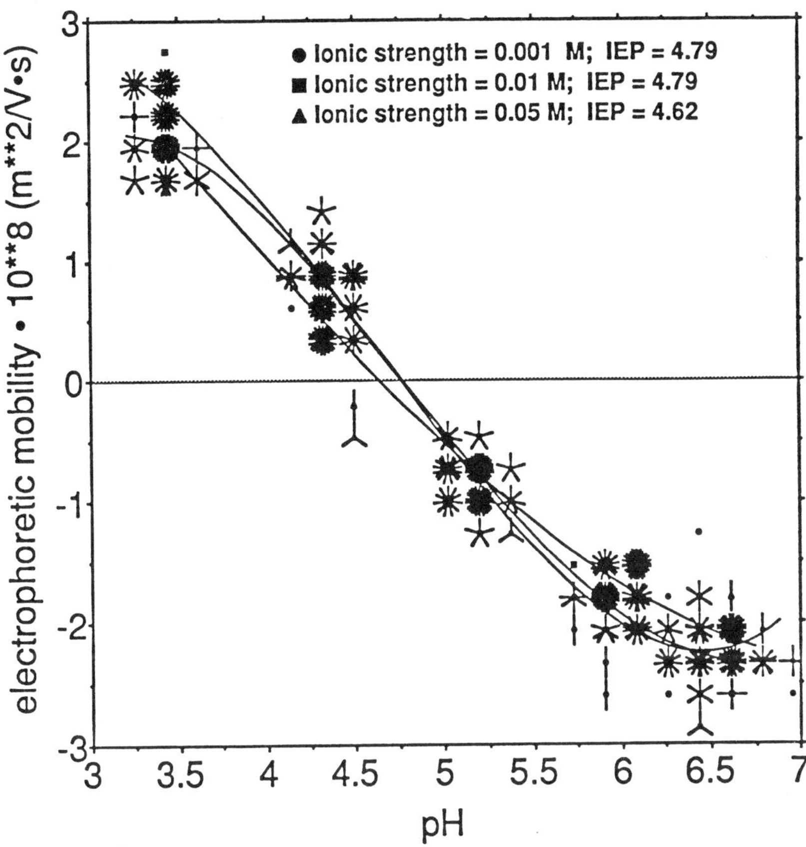

Figure 3. IEP values determined from each separate mobility measurement, with the data displayed using variable cellulation.

Before phosphated goethite can be used as a standard material, further studies are needed. While this comparison of the performance of three instruments suggests that these units provide similar readings on acidic suspensions, a more extensive round robin with more participants is needed to fully justify this conclusion. Such a round robin must include well-defined protocols for obtaining measurements and, if performed with phosphated goethite, should employ a higher solids loading than used in this study. (We recommend 250 mg solid/L or 0.006 vol%.) Further studies are also needed to determine if goethite can be synthesized in a reproducible manner by different groups, to determine the aging behavior of freshly synthesized goethite (i.e., the hydration time needed for the pH and mobilities of goethite suspensions to stabilize), and to determine the shelf life of aqueous suspensions of phosphated goethite. It would also be useful to check the performance of these instruments under alkaline conditions with a similar round robin, if a suitable material can be located. We have performed such a study with aqueous suspensions of silicon nitride, but silicon nitride is too reactive to serve as a useful standard.

While aqueous suspensions of latex particles may be suitable standards for calibrating mobility measurements under specific conditions, such materials are not really appropriate for checking IEP measurements in aqueous suspensions. For this purpose, one would prefer to use a material that displays an obvious IEP without showing severe aging effects, such as an appropriate hydrous oxide.

ACKNOWLEDGEMENTS

This research was sponsored by the U.S. Department of Energy, Assistant Secretary for Conservation and Renewable Energy, Office of Transportation Technologies, as part of the Ceramic Technology Project of the Materials Development Program, under contract DE-AC05-84OR21400 with Martin Marietta Energy Systems, Inc.

REFERENCES

1. R.J. Hunter, <u>Zeta Potential in Colloid Science: Principles and Applications</u>, Academic Press, New York, NY, 1981.

2. J.-F. Wang, R.E. Riman and D.J. Shanefield, "Reliable Electrophoretic Mobility Measurement for Ceramic Powders", 240-246, <u>Ceramic Transactions</u>, Vol. 26, American Ceramic Society, Westerville, OH, 1992.

3. W. Stöber, A. Fink and E. Bohn, "Controlled Growth of Monodisperse Silica Spheres in the Micron Size Range", J. Colloid Interface Sci., 26, 62-69, 1968.

4. D.D. Hansmann and M.A. Anderson, "Using Electrophoresis in Modeling Sulfate, Selenite, and Phosphate Adsorption onto Goethite", Environ. Sci. Technol., 19, 544-551, 1985.

5. W.A. Zeltner and M.A. Anderson, "Surface Charge Development at the Goethite/Aqueous Solution Interface: Effects of CO_2 Adsorption", Langmuir, 4, 469-474, 1988.

6. M.I. Tejedor-Tejedor and M.A. Anderson, "Protonation of Phosphate on the Surface of Goethite As Studied by CIR-FTIR and Electrophoretic Mobility", Langmuir, 6, 602-611, 1990.

7. M.L. Machesky and M.A. Anderson, "Calorimetric Acid-Base Titrations of Aqueous Goethite and Rutile Suspensions", Langmuir, 2, 582-587, 1986.

8. R.W. O'Brien and L.R. White, "Electrophoretic Mobility of a Spherical Colloidal Particle", J. Chem. Soc. Faraday Trans. II, 74, 1607-1626, 1978.

9. R.J. Atkinson, A.M. Posner and J.P. Quirk, "Crystal Nucleation in Fe(III) Solutions and Hydroxide Gels", J. Inorg. Nucl. Chem., 30, 2371-2381, 1968.

10. R.J. Hunter, <u>Foundations of Colloid Science</u>, Vol. 1, Chap. 6, Clarendon Press, Oxford, UK, 1987.

ELECTROKINETIC CHARACTERISTICS OF Pb(Ti,Zr)O₃ AND ITS OXIDE CONSTITUENTS

Jasenka Biscan, Ruder Boskovic Institute, Zagreb, Bijenicka 54, Croatia
Marija Kosec, Jozef Stefan Institute, Ljubljana, Jamova 39, Slovenia

ABSTRACT

$Pb(Zr,Ti)O_3$ solid solutions were synthesized by solid state reaction of the constituent oxides. Microelectrophoresis was applied to analyze the isoelectric conditions of the individual oxide components and of the synthesized complex oxides. According to the similarity between the isoelectric points of ZrO_2 and PZT, it seems that zirconium oxide has a major influence on the electrokinetic charge and potential of the synthesized complex oxide (PZT).

INTRODUCTION

It is becoming clear that the stability of the suspension in various steps of preparation of ceramic materials is of critical importance (1). This in most cases is determined by the electrostatic particle-particle interactions effected by the surface charge. A semi-direct method for the determination of the surface or interfacial charge and potential is the measurement of the electrophoretic mobilities of dispersions under predetermined conditions of the surrounding solution, such as pH, ionic strength, and species composition (2). $Pb(Zr,Ti)O_3$ (PZT) is a well known piezoelectric ceramic material (3). The present study deals with the behavior of the oxides PbO, ZrO_2 and TiO_2 and PZT of general formula $PbZr_{0.52}Ti_{0.48}O_3$ synthesized at a temperature of 800-1200°C by solid state reaction. The electrokinetic characteristics of PZT and its oxide constituents in aqueous solution of different composition, ionic strength and pH could help us to define the optimal conditions for the synthesis of well reacted PZT powders, and could help in preparing a stable PZT suspension for further processing, including spray drying, tape casting, slip casting, etc.

The complex oxides synthesized were analyzed with respect to stoichiometry (deficiency or excess of PbO) and synthesis temperature. Stoichiometric PZT is difficult to obtain because of the high vapor pressure of PbO and its loss during sintering. Therefore the surface properties of nonstoichiometric

PZT powders might be interesting. Common synthesis temperatures of PZT are between 800° and 1000°C (4). A temperature of 1200°C was selected to prepare a well reacted powder.

The crystal structure of PZT is a function of the Zr/Ti ratio. The investigated materials are characterized by the morphotropic phase boundary and in accordance with literature data are mixtures of tetragonal and rhombohedral phases (3). XRD spectra of samples synthesized at lower temperature show that solid solution formation is incomplete. From these spectra a series of solid solutions based on a Zr-rich rhombohedral structure and a Ti-rich tetragonal structure can be identified. The results are in agreement with published data concerning the mechanism of PZT solid solution formation (5-8).

Composition	Synthesis Temperature (°C)	Crystal Structure	Particle Size (μm)	Agglomerate Size (μm)
PbO		litharge massicot	2	10-25
TiO_2		anatase	0.2	10-20
ZrO_2		baddeleyite	0.4	10-15
PZT_{800}	800	PZT (r,t)	1	15
PZT_{900}	900	PZT (t,r)	1	10-15
PZT_{950}	950	PZT (t,r)	1	10-15
PZT_{1000}	1000	PZT (t,r)	1-3	10-15
PZT_{1200}	1200	PZT (t), and minor amount PZT (r)	1-5	40
$P_{0.99}ZT$	900	PZT (r,t)	1	40-60
$P_{1.025}ZT$	900	PZT (t,r) PbO massicot	0.5-3	10-20

t - tetragonal, r - rhombohedral

Table 1. Characteristics of Simple and Complex Oxides

The estimated particle size of PZT powders ranges from 1 to several micrometers. A slight coarsening of particles of PZT powders with an excess of PbO and those synthesized at higher temperatures was observed. The estimated agglomerate sizes were 10 μm to 100 μm.

The specific surface areas of oxides were 0.44 m^2/g for PbO, 3.4 m^2/g for ZrO_2 and 8.4 m^2/g for TiO_2. PZT surface area varied between 0.12 to 0.34 m^2/g, depending on the synthesis temperature.

Electrokinetics

A suspension of PbO either in redistilled water or in an electrolyte (NaCl or NH_4Ac) is characterized by a high pH of 9.9 $\pm$ 0.1 and a high positive zeta potential. At a very low suspension concentration of less than 10 mg/100 cm^3 there is a strong decrease of zeta potential toward zero or even a negative value. This is due to non-equilibrium conditions and the effect of specific adsorption of anions (Cl^-). However, this fact is not particularly important for the discussion on ceramic suspensions. Unlike in the higher electrolyte concentrations, the system is fairly stable in 10^{-4} M NaCl showing a zeta potential of +50 $\pm$ 2 mV, even after 48h equilibration.

The dispersion of PbO in ammonium acetate solution shows a high stable positive zeta potential of +51 $\pm$ 1 mV at three initial electrolyte concentrations of NH_4Ac. Visual inspection of the suspension in the electrophoretic cell showed a stable homogeneous suspension of tiny particles.

The titration of PbO dispersions show different characteristics for the two electrolytes. The dispersions in 10^{-4} M NaCl solutions become unstable and negatively charged only at very high pH with the isoelectric point at about pH 11. Down to pH 5 a highly stable positive zeta potential was observed. The acetate suspensions are very stable, but only in the range of pH 6-9. Above pH 9 large aggregates form (precipitation of hydroxides is also possible), and below pH 6 strong progressive dissolution occurs. The titration curves for PbO in NaCl and NH_4Ac together with the results for other oxides are shown in Figures 1 and 2.

The suspension of TiO_2 in NaCl gave a pH of 6.2. All suspensions are very stable and the mobility distribution peaks are sharp, indicating a homogeneous sample. Some increase of zeta potential with suspension concentration was found. However, stable conditions were reached at a suspension concentration as low as 20 mg/100 cm^3. The isoelectric point in NaCl solution is at pH 3.5 and close to pH 3 in NH_4Ac solution.

A suspension of ZrO_2 in 10^{-4} M NaCl or NH_4Ac gives a pH of 6.0-6.3. The isoelectric point under equilibrium conditions is at pH 6.8 in NaCl and about pH 7.5 in NH_4Ac. Therefore this pH region could be assigned as a low stability region for this system. A rather strong dependence of zeta potential on the suspension concentration in the case of NaCl as electrolyte was found.

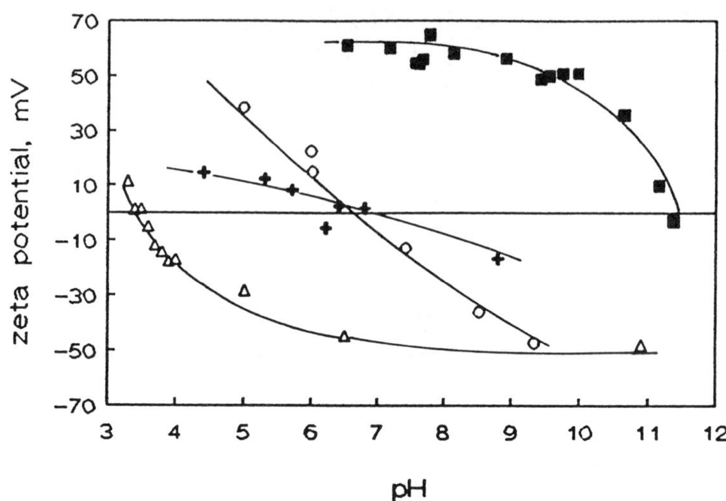

Figure 1. Titration and zeta potential of PZT and composite oxides in 10^{-4} M NaCl. (•) PbO, (Δ) TiO$_2$, (o) ZrO$_2$, (+) PZT.

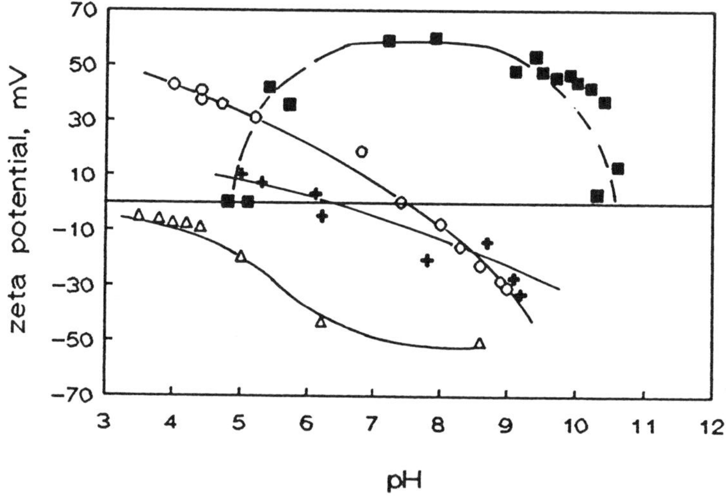

Figure 2. Titration and zeta potential of PZT and composite oxides in 10^{-4} M NH$_4$Ac. (•) PbO, (Δ) TiO$_2$, (o) ZrO$_2$, (+) PZT.

     Characterization Techniques for the Solid-Solution Interface

Despite the fact that the behavior of individual oxide components are somewhat different in the two electrolytes investigated, the main results for PZT are very similar for both chloride and acetate systems. The similarity between the isoelectric points for PZT and ZrO_2 suggests the dominant role of ZrO_2 in charging the complex oxide in NaCl media. In acetate, the isoelectric point of PZT differs significantly from that of ZrO_2. The specific adsorption of acetate ions here plays a significant role.

With respect to the synthesis temperature, there is no clear indication of any "critical temperature" indicating the regions of partial and full reaction between components. Statistically, a relative impact of positively and negatively charged particles gives an average result for each case.

The different concentrations of PbO in PZT did not cause any significant changes compared to stoichiometric PZT. It seems that the real influence of nonstoichiometric conditions are somehow compensated for by surface impurities.

In Figures 3 and 4 the graphs show the zeta potential of composite oxides and PZT under selected pH conditions: 5, 7, 9 and 11. Zeta potential values are the lowest for PZT in the entire pH range. This indicates internal neutralization of oxides where surface charges are significantly compensated or blocked. It seems that primarily a neutralization occurs between PbO and TiO_2, whereas most of the active sites of ZrO_2 remain exposed. One can speculate that the results are in agreement with published data showing that ZrO_2 particles are the last reacting species in the reaction between PbO, TiO_2 and ZrO_2 giving PZT (6-8).

CONCLUSION

In conclusion, at the present stage of the investigations, it is evident that many problems concerning the electrokinetic behavior of PZT's and their constituent oxides are highlighted but not resolved. The pH and electrolyte concentration regions were identified, indicating the conditions for possible good reactivity between the oxide components. Apart from stability regions, the electrolytes used in this work do not behave as "indifferent" species. The isoelectric region for PZT is very close to the isoelectric region for ZrO_2. The present work does not indicate any remarkable differences in the behavior of PZT samples of different composition and synthesized at various temperatures. The real characteristics of the surfaces are somehow compensated for by surface impurities. Finally, each of the problems mentioned deserves a separate, more detailed investigation.

ACKNOWLEDGMENT

The work was performed under the project NIST 981. The financial support from the National Institute for Standards and Technology, USA, the

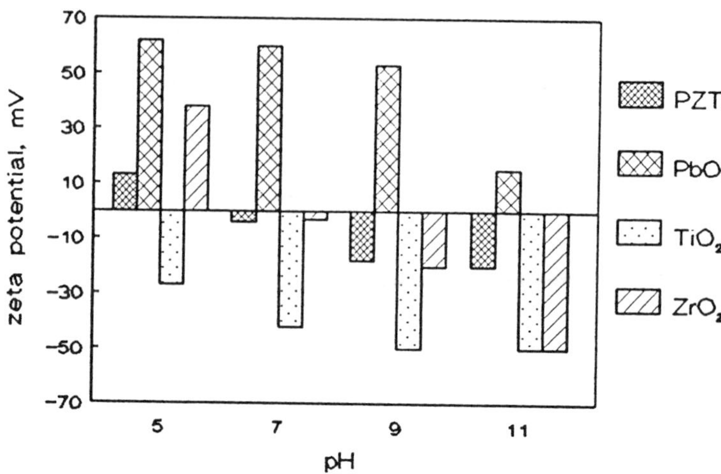

Figure 3. Estimated zeta potential of PZT and its oxide constituents in 10^{-4} M NaCl.

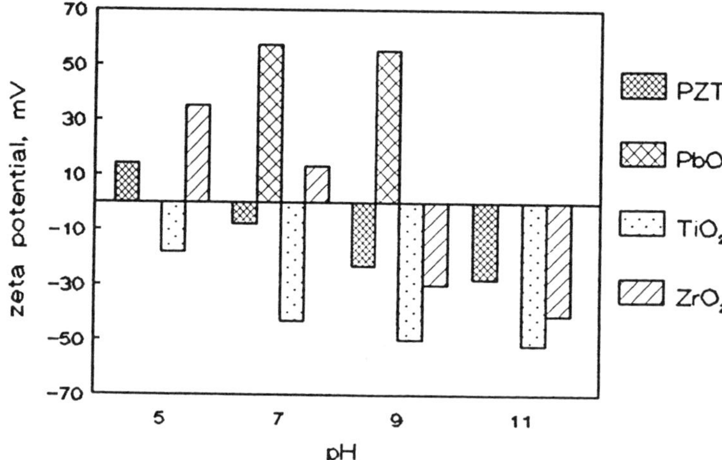

Figure 4. Estimated zeta potential of PZT and its oxide constituents in 10^{-4} M NH_4Ac.

   Characterization Techniques for the Solid-Solution Interface

Ministry of Science and Technology of Croatia and the Ministry of Science and Technology of Slovenia is gratefully acknowledged.

REFERENCES

1. F. Aldinger and H-J. Kalz, "The Importance of Chemistry in the Development of High-Performance Ceramics," *Angew. Chem. Int. Ed.,* **26**, 371-381 (1987).
2. E. Matijevic, "Colloid Science in Ceramic Powders Preparation," *High Technology Ceramics* (P. Vincenzini Ed.,), Elsevier, Amsterdam, 1987.
3. B. Jaffe, W.R. Cook, H. Jr. Jaffe, *Piezoelectric Ceramics,* Academic Press, London, 1970.
4. D.A. Buckner and P.D. Wilcox, "Effect of Calcining on Sintering of Lead Zirconate-Titanate Ceramics," *Ceram. Bull.,* **51**, 218-222 (1972).
5. M. Kosec and D. Kolar, "Effect on the Reactant Chemical Composition and Atmosphere on PZT Solid Solution Formation," *La Ceramica,* 26-32 (1985).
6. B.V. Hiremath, A.I. Kingon and J.V. Biggers, "Reaction Sequence in the Formation of Lead Zirconate-Lead Titanate Solid Solutions: Role of Raw Materials," *J. Am. Ceram. Soc.,* **66**, 790-793 (1983).
7. S. Venkataramani and J.V. Biggers, "Reactivity of Zirconia in Calcining of Lead Zirconate-Lead Titanate: Compositions Prepared from Mixed Oxides," *Ceram. Bull.,* **59**, 462-466 (1980).
8. A.I. Kingon, P.J. Terblanche and J.B. Clark, "Effect of Reactant Dispersion on Formation of PZT Solid Solution," *Ceram. Int.,* **8**, 108-113 (1982).

DETERMINATION OF PARTICLE PROPERTIES FROM ELECTROACOUSTIC MEASUREMENTS

R.W. O'Brien, School of Chemistry, The University of Sydney, NSW 2006, Australia

PARTICLE CHARGE

In this paper I will discuss effects involving alternating electric fields and sound waves in suspensions of small particles in a liquid. Such suspensions are called "colloidal sols."

When an alternating electric field is applied to a colloid, sound waves are generated. Conversely, when an ultrasonic beam is directed into a colloid, alternating electric fields and currents are set up. Phenomena such as these, involving either the generation of sound waves by electric fields or the generation of electric fields and currents by soundwaves are called electroacoustic effects. In this paper I will describe the mechanisms that give rise to these effects and show how electroacoustic measurements can be used to determine properties of the colloidal particles.

Electroacoustic effects occur because the particles are electrically charged. This charge may arise in a number of ways, such as the dissociation of surface groups or the preferential desorption or adsorption of ions at the surface. In nearly every case the electric charge lies on the particle surface rather than the interior of the particle.

This charge can be controlled by altering the concentration of "potential determining ions" in the suspending electrolyte. In those colloids where the particles have acids or base groups on the surface, the hydrogen ion is potential determining, and the charge is controlled by the pH.

The particle charge affects most of the bulk properties of the colloid. For example, in many colloids the electrical repulsive forces between the particles stop them from coagulating under the influence of the van der Waals attractive force, and thus such suspensions can be made to coagulate, or can be rendered stable by controlling the concentration of potential determining ions.

The charge on the particles is balanced by an equal and opposite charge on the ions in the liquid. These mobile charges form a diffuse cloud around the particle. This diffuse layer and the layer of charge at the particle surface are known as the electrical double layer. The thickness of the diffuse layer is determined by the balance between the electrical forces, which try to pull the ions onto the surface, and the thermal forces which try to spread the ions uniformly. It can be shown (1,p332) that the charge density in the diffuse layer decreases exponentially with distance from the particle surface. The decay length is denoted by the symbol κ^{-1}, and is called the Debye length, or simply the double-layer thickness. In a symmetric two-species electrolyte (ibid)

$$\kappa^{-1} = 3.2888\sqrt{c} \ (nm^{-1})$$

where c is the ionic strength in mol L^{-1}. Thus for a 0.001M electrolyte the double layer is about 10 nm thick, and for 0.1M it is 1nm.

There is much more that can be said on this important subject of particle charge, but for the purposes of this article this brief introduction should suffice. Readers who wish to know more on this subject should consult references 1 and 2.

ESA AND CVP EFFECTS

As mentioned above there are two types of electroacoustic effects: those involving the generation of sound waves by electric fields, and those involving the generation of electric currents or voltages by applied sound waves. In this section I will describe the physical mechanisms that give rise to these effects.

Consider first the generation of sound waves by electric fields. This phenomenon is called the electrokinetic sonic amplitude, or ESA effect. It was only discovered in the early 1980's (3).

The ESA effect occurs because the particles are electrically charged. Thus when an alternating voltage is applied to the colloid the particles move, first towards one electrode and then the other. As they move, they create pressure disturbances in the surrounding liquid. These disturbances propagate out from the particle surface in the form of sound waves. The ESA effect is just the superposition of sound from the individual particles.

The other class of electroacoustic effect involves the generation of electric fields and currents by applied sound waves. The sound waves cause the particles and liquid to move backwards and forwards. If the particle density is different from the liquid there will be a relative motion between the two phases. Thus, relative to the particle, the liquid will sweep one way and then the other across the particle surface at the frequency of the sound wave. As the liquid sweeps over the surface it carries double layer ions with it. Thus the centre of charge of the double layer is displaced from the particle centre of charge. Each particle and its double layer act like an alternating dipole. The particle dipole fields sum to produce macroscopic electric fields. The open-circuit voltage measured between any two points in the sound wave field is called the colloid vibration potential, or CVP.

This effect was predicted for ionic solutions by Debye over half a century ago (4) and extended to colloids by Hermans in 1938 (5).

    Characterization Techniques for the Solid-Solution Interface

Note that the CVP effect requires a density difference between the particle and the liquid. If the particle is neutrally buoyant it will move in the sound wave in the same way as the liquid. Thus there will be no relative motion, and hence no CVP in a suspension of neutrally buoyant particles.

SIZE OF ELECTROACOUSTIC EFFECTS

Both the ESA and CVP are small effects. A suspension of particles with a density of 2g cm^{-3} and a particle volume fraction of 0.05 will typically generate an ESA sound wave with a pressure amplitude of 10^{-4} atmospheres in an electric field of 100 V/cm. If the CVP effect is generated in the same suspension by an ultrasonic beam with a pressure amplitude of 0.1 atmospheres, voltage differences of about 1mV are set up.

It was probably because of the difficulties involved in measuring such small signals that the science of electroacoustics advanced so slowly after the initial discovery of the CVP effect. During the last decade however, commercial devices were developed which accurately measure these signals. This has sparked renewed interest in the area and led to some major developments in the science of electroacoustics.

APPLICATION OF ELECTROACOUSTIC MEASUREMENTS

Later in this paper I will show that the electroacoustic signal is related to the size and charge of the colloidal particles. Thus electroacoustic measurements can be used for obtaining information on particle size and charge.

These are probably the two most important particle properties, in that they influence so many bulk properties of the colloid. For example, the opacity of paints, inks and other coatings are strongly influenced by particle size, and the strength of a ceramic is also dependent on particle size. Particle charge influences the flow properties and the stability of a suspension to flocculation.

Although there are already a number of techniques for determining particle size and charge, they are nearly all limited to very dilute suspensions, for they involve the passage of light through the suspension, and so the suspension must be optically clear. The light scattering technique for example, requires particle volume fractions around 0.01% or less. Electroacoustic effects can be measured in suspensions of arbitrary concentration, and thus there is the possibility that these measurements may allow the determination of size and charge in concentrated suspensions. This would allow the *in situ* determination of size and charge in colloidal process streams and thereby facilitate process control operations.

For the past five years the colloids group at the University of Sydney has been involved with Matec Applied Sciences, an American company, in the development of a commercial electroacoustic device for measuring size and charge in concentrated suspensions. That device has now been commercially released in the United States. It is known as the AcoustoSizer.

MEASUREMENT OF ELECTROACOUSTIC EFFECTS

Although this is not a paper on the AcoustoSizer, there are some important features of electroacoustics that can be learned from a study of the measurement technique that is used in this device.

In Figure 1 we show a sketch of the AcoustoSizer cell. The suspension is contained in a cup, in contact with a pair of disc shaped electrodes. The electrodes are attached to glass rods which pass through sealed holes on the sides of the cup.

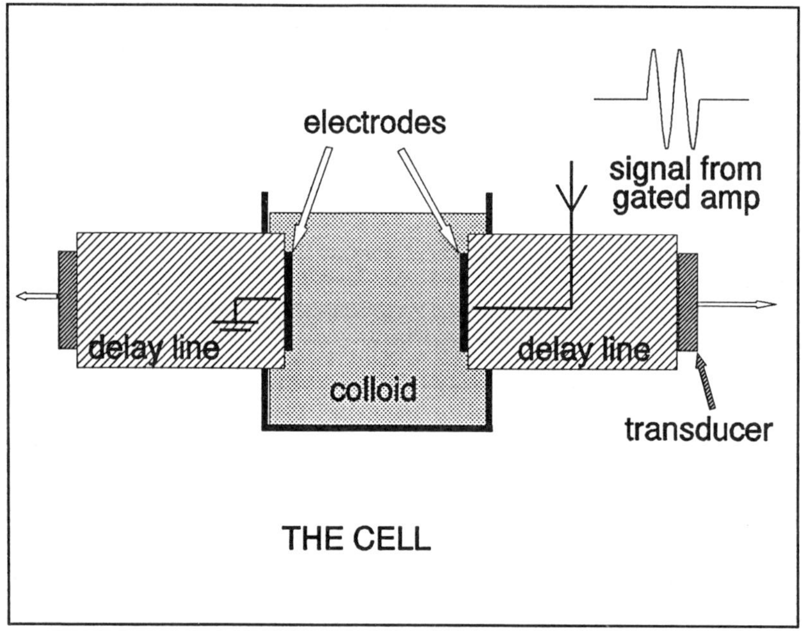

Figure 1. The AcoustoSizer cell

     Characterization Techniques for the Solid-Solution Interface

The device is usually operated in ESA mode, because the ESA signal has a better signal to noise ratio than the CVP in aqueous systems, for reasons that will be described in the following section. The ESA effect is generated by an alternating voltage difference applied across the electrodes. Measurements are made over a range of frequencies between 300kHz and 12MHz. The ESA sound waves travel down the glass rods where they are detected by a piezoelectric transducer.

Although these sound waves arise from the particles, the resulting macroscopic sound wave appears to emanate from the electrodes. This occurs because of an asymmetry in the particle sound wave fields. For a particle oscillating in an infinite liquid the pressure at a point on the upstream side is equal and opposite to the pressure at an equidistant point on the downstream side. In other words, for each crest in the sound wave on the upstream side (the side to which the particle is moving) there is a corresponding trough at the same distance on the downstream side. Thus there can be no ESA effect in an infinite suspension, for the troughs from all the particles on the right of any point will be cancelled by the crests of the particles on the left. In a bounded suspension this is also true for interior points which lie more than a distance ct from the boundary, where c is the sound speed and t is the time after the applied voltage is turned on. These points are not affected by the boundaries and are effectively in an infinite suspension. For those particles within a distance ct of the boundary, the cancellation of the sound waves is incomplete and thus macroscopic sound waves appear in these regions.

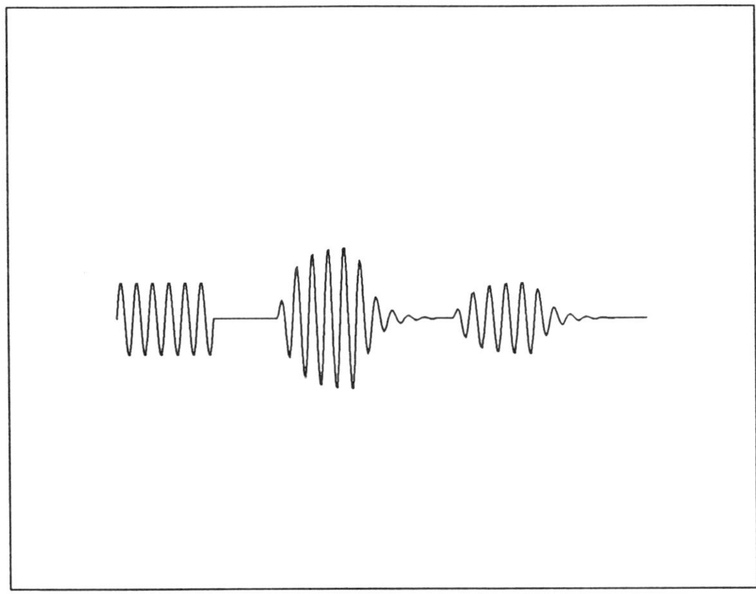

Figure 2. The ESA signal

Figure 2 shows a sketch of the voltage signal from the transducer as a function of time. The applied voltage is a pulsed sinusoid, that is a sinusoid which is only turned on for a short time. The first signal on the diagram occurs at the time that the voltage is applied across the colloid. This signal has nothing to do with the ESA effect, for there has not been time for the sound to travel down the rod. It is electrical "cross talk," an unwanted signal radiated from the high voltage pulse across the cell and picked up by the transducer electrodes. It is to avoid this noise that the glass rods are used; the applied pulse is turned off before the sound wave has reached the transducer.

The first pulse after the cross-talk comes from the sound wave "generated" by the nearest electrode. After that comes another pulse which emanates from the other electrodes.

In the AcoustoSizer the signal processing scheme determines the Fourier Transform of thee first pulse after the x-talk. In other words the electronics extracts from the pulse the sinusoidal Fourier component of the pulse at the frequency of the applied voltage. We do this to facilitate comparison with theory, which assumes sinusoidal disturbances.

Although this, device is designed to measure ESA it could also be used to measure the CVP. In this case the alternating voltage pulse is applied to the transducer at the end of the rod, instead of across the suspension. This causes the transducer to expand and contract at the frequency of the applied voltage, generating sound waves which travel down the rod. As the sound wave pulse crosses the nearer electrode and passes into the suspension, a CVP voltage pulse is generated between the electrodes. In this case the electrodes *detect* the signal rather than generating it. Another CVP pulse appears when the sound wave reaches the electrode on the other side of the colloid.

RECIPROCAL RELATION BETWEEN ESA AND CVP

Although the ESA and CVP effects arise in different ways, one from particle motion and the other from double layer distortion, it can be shown (6) that the two are linked by the simple reciprocal relation

$$V_{ESA} = Z \ V_{CVP}$$

Here V_{ESA} is the (Fourier transform of the) voltage generated across the transducer, divided by the voltage applied across the cell to generate that signal, and V_{CVP} is the CVP voltage generated by the sound wave as it passes through the colloid, divided by the voltage applied across the transducer in order to generate that sound wave. Z is the electrical impedance of the transducer divided by the cell impedance.

From this relation it can be seen that the ESA voltage ratio is greater than the CVP when the magnitude of Z is greater than one. For aqueous suspensions this is

                   Characterization Techniques for the Solid-Solution Interface

always the case and hence the ESA measurement has a greater signal to noise ratio than the CVP.

In our discussion of the CVP we pointed out that the effect requires a density difference between the particles and the solvent. From the above reciprocal relation it follows that this must also be true for the ESA effect. Thus, if an electric field is applied to a suspension of neutrally buoyant particles no macroscopic sound will be generated, even though the particles are oscillating back and forth. This is because the ESA sound wave is affected not just by the particle motion, but also by the force which the applied field exerts on the diffuse charge cloud surrounding the particle. That force also generates a soundwave, and that wave tends to cancel the wave from the particle, for the particle and diffuse layer exert equal and opposite forces on the liquid. For a neutrally buoyant particle this cancellation of the two soundwaves is complete.

DYNAMIC MOBILITY

We now turn to the problem of determining the particle size and charge from electroacoustic measurements.

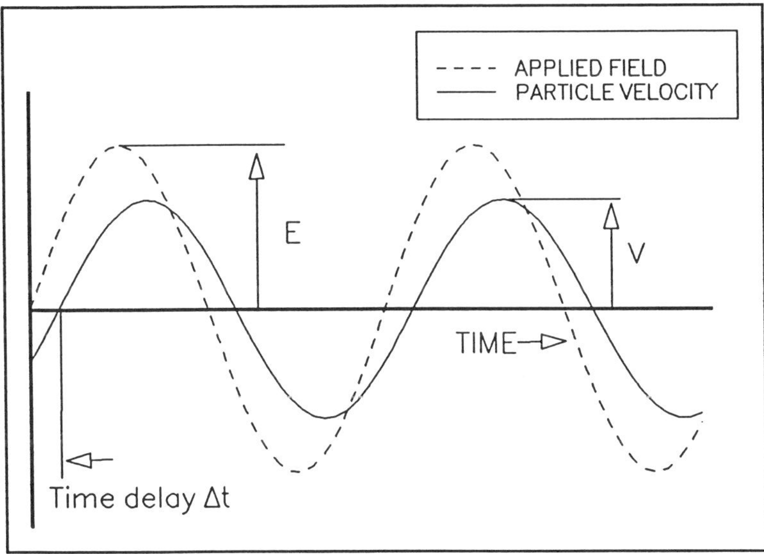

Figure 3. Particle motion in an alternating electric field

This can be thought of as a two-step process: in the first step we calculate a particle quantity known as the dynamic mobility from the ESA signal. In the second step we determine size and charge from that mobility.

In this section I will describe the first step, beginning with the definition of the dynamic mobility.

Suppose that a suspension is subjected to an alternating electric field. After a microsecond or so the particles will move backwards and forwards in a sinusoidal fashion. If the frequency of the field is sufficiently high, inertia forces will cause the particle to lag the applied field. This is illustrated in Figure 3, where the broken line represents the applied field strength and the unbroken curve is the particle velocity.

The amplitude of the particle velocity V is proportional to the amplitude of the applied field. Thus the ratio V/E is independent of field strength and is a particle property. The time delay Δt between the field and the particle motion is also independent of field strength. By definition, the dynamic mobility μ is a complex quantity with magnitude V/E and argument $2\pi f(\Delta t)$, where f is the applied frequency in MHz and Δt is measured in seconds.

As the applied frequency tends to zero the time delay vanishes and hence μ becomes a real quantity, equal to the familiar electrophoretic mobility defined for a static applied field.

The formula for determining the dynamic mobility from the ESA involves a quantity Z^a known as the Acoustic Impedance of the suspension. This is defined as follows: If a flat surface in contact with the suspension is made to oscillate backwards and forwards with a (complex) velocity v, then the force per unit area required to bring about this motion is $Z^a v$.

Now for the formula linking ESA to μ. From the solution of the electroacoustic equations for a concentrated suspension (6) it can be shown that the ESA pressure wave in the AcoustoSizer delay rod is given by

$$\phi \frac{\Delta \rho}{\rho} \mu E \frac{Z^a}{Z^a + Z_r^s} \qquad [1]$$

for suspensions of arbitrary concentration, where ϕ is the particle volume fraction, ρ is the liquid density and $\rho + \Delta \rho$ is the particle density. E is the field strength on the suspension side of the electrode and Z_r^a is the acoustic impedance of the rod.

In most applications μ and Z^a are unknowns. In order to determine either quantity we need another measurement. The second delay rod in the AcoustoSizer, shown on the left in figure 1 is designed to make an acoustic measurement of Z^a, thereby facilitating the determination of μ.

There is one other aspect to this procedure that should be mentioned. Z^a is very sensitive to bubbles in the suspension, much more so than μ. This is because

 Characterization Techniques for the Solid-Solution Interface

bubbles are very efficient absorbers of sound energy. Thus by extracting the dynamic mobility from the ESA we greatly reduce our sensitivity to bubbles.

From the formula (1) it can be seen that the ESA increases with $\phi\Delta\rho$. For very dilute suspensions, or suspensions with small $\Delta\rho$, the signal may be too small to measure. In the AcoustoSizer the lower limit for ϕ is around 0.005 if $\Delta\rho/\rho$ is 1.

FORMULAE FOR THE DYNAMIC MOBILITY

The second step is the determination of size and charge from the dynamic mobility. This involves the use of theoretical formulae for μ.

The only formulae that have appeared in the literature so far are limited to particle volume fractions of less than 5% or so. In the derivation of these formulae it is assumed that there is a smooth surface dividing the particle and any adsorbed material from the surrounding liquid. The equilibrium voltage difference between this surface and the electrolyte beyond the double layer called the ζ-potential (2). Since the true particle surface will not be molecularly smooth, we must view ζ as an effective surface potential, representing some sort of average over the true surface. Despite the model-dependent nature of this definition, the ζ potential is an important colloid property in that it is directly linked to properties such as suspension viscosity and stability (2).

For spherical particles with thin double layers the formula for the dynamic mobility takes the form (7)

$$\mu = \frac{\epsilon\zeta}{\eta} G(\frac{\omega a^2}{\nu}) \qquad [2]$$

where ϵ and η are the solvent permittivity and viscosity respectively. The symbol a is the particle radius, ω is the angular frequency of the applied field and ν ($=\eta/\rho$) is the kinematic viscosity of the solvent.

The quantity G represents the effect of inertia forces. At low frequencies, that is when $\omega a^2/\nu$ is much less than 1, G=1 and the above formula reduces to the well-known Smoluchowski formula for the electrophoretic mobility. At these frequencies μ depends on ζ, but not on particle size.

As $\omega a^2/\nu$ approaches unity the inertia forces become significant and as a result the magnitude of μ drops and the particle develops a phase lag. Both the magnitude and phase lag depend on particle size at these frequencies.

As $\omega a^2/\nu$ continues to increase the magnitude of μ diminishes until eventually the ESA signal is too small to measure.

From this discussion it can be seen that if $\omega a^2/\nu$ is either too small or too large it is not possible to assess particle size from the ESA measurement. Hence the range of particle sizes that can be determined from ESA measurements is limited by the frequency range. The frequency range of the AcoustoSizer has been chosen to allow sizing of particles in the range 0.1 to 10 microns. For smaller particles ζ can be determined, but the inertia forces are too small to allow sizing of these particles.

The formula (2) for dynamic mobility has recently been extended to the case of spheroidal particles of arbitrary aspect ratio (8). The main result of that analysis is that the dynamic mobility spectrum for a spheroidal particle has a very similar shape to that of a sphere with an "effective" radius that depends on the aspect ratio and density of the particle. Thus it would be very difficult to distinguish a spheroid from a sphere using electroacoustic data. On the other hand this means that the theory for spheres can be used for determining the equivalent particle radius of spheroidal particles as well.

Work is presently underway on the calculation of μ in concentrated suspensions of spherical particles.

EXPERIMENTAL RESULTS

In Figure 4 we show size distributions obtained on two BCR quartz suspensions. These suspensions are prepared as sizing standards by the Community Bureau of Reference in Brussels. They come supplied with size data obtained by manual sedimentation.

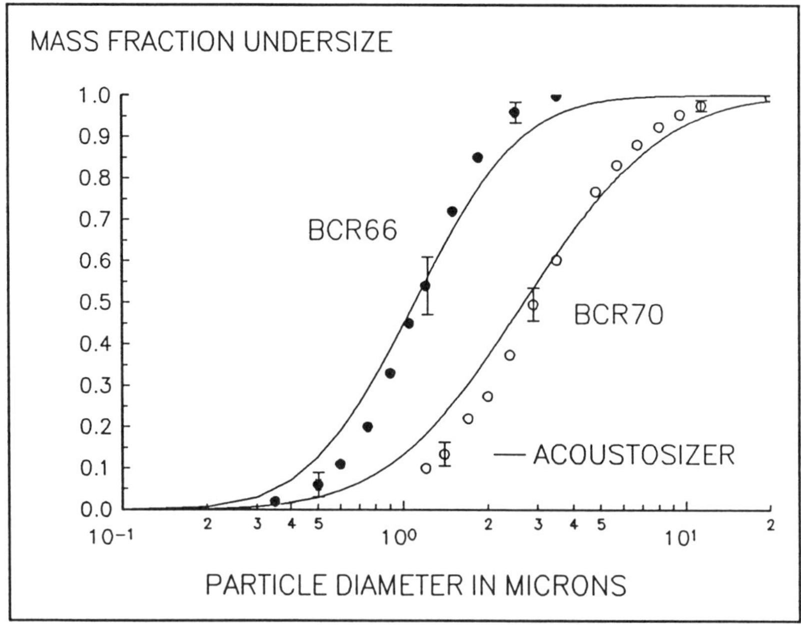

Figure 4. Particle size distributions of BCR66 and BCR70 quartz suspensions.

　Characterization Techniques for the Solid-Solution Interface

The points represent the sedimentation measurements and the curves are the distributions obtained on the AcoustoSizer. At present the sizing algorithm assumes a log-normal distribution. The median diameter, the standard deviation, and the ζ potential are adjusted to give the best fit between the measured and theoretical mobility spectra.

The AcoustoSizer and sedimentation data are in reasonable agreement. The discrepancies could be due to the fact that the effective spherical radius in sedimentation is not the same as the effective ESA radius.

In Table 1, ζ potentials obtained from the AcoustoSizer are compared with the values on the Rank electrophoresis device and the Malvern ZetaSizer.

Colloid	pH	Acoustophoresis	Rank	Zetasizer
TiO_2	4.0	37	38	
	8.2	-6	-6	
	9.0	-20	-22	
Si_3N_4	3.5	26	24	
	8.9	-39	-43	
Kaolin	4.4	-28	-26	
	9.8	-37	-43	
Alumina	5.0	47	58	53
Si_3N_4	10.0	-35	-42	-51
TiO_2	4.0	40	50	57

Table 1. Comparison of Zeta potentials (in millivolts) obtained by three different techniques

Both the Rank and the Malvern require very dilute suspensions. This dilution can alter the zeta potential. For example if the true background electrolyte is not known or if it depends on adsorption processes on the particle then the dilution will alter the electrolyte and so change zeta. Furthermore for the diluted sols the particle surface is so small that trace amounts of impurity in the solution can adsorb on the surface and significantly alter zeta potential. These effects could be responsible for the deviations in the zetas obtained with the three devices.

As I mentioned in section 4 it is one of the great advantages of electroacoustic measurements that they can be made without dilution. With this technique it should be possible to explore new areas of colloid science involving particle interactions and to study *in situ* the concentrated suspensions that occur in industrial applications.

ACKNOWLEDGMENT

The work that is described here is the result of a joint effort. In particular I want to acknowledge the role of David Cannon of Matec Applied Sciences who has been working closely with me for the past five years and has made major contributions to the design and testing of the AcoustoSizer. We have also been assisted by Profs. R.J. Hunter and J.K. Beattie at the School of Chemistry at the University of Sydney.

REFERENCES

1.	Hunter, R.J., "Foundations of Colloid Science," vol. 1, Clarendon Press, Oxford, 1987.
2.	Hunter, R.J., "Zeta Potential in Colloid Science," Academic Press, New York, 1981.
3.	Oja, T., Petersen, G.L., and Cannon, D.W., United States Patent no. 4,497,208 (1985).
4.	Debye, P., *J. Chem. Phys.,* **1**, 13 (1933).
5.	Hermans, J., *Philod. Mag.,* **25**, 426 (1938).
6.	O'Brien, R.W., *J. Fluid Mech.,* **212**, 81 (1990).
7.	O'Brien, R.W., *J. Fluid Mech.,* **19O**, 71 (1988).
8.	Loewenberg, M., and O'Brien, R.W., *J. Colloid Interface Sci.,* **150**, 158 (1992).

          Characterization Techniques for the Solid-Solution Interface

THE IMPLEMENTATION OF ACOUSTOPHORESIS FOR THE CHARACTERIZATION OF SILICON NITRIDE BASED AQUEOUS SUSPENSIONS

J.P. Pollinger, D.D. Newson, and J.J. Nick
Garrett Ceramic Components
Allied-Signal Aerospace Company
Torrance, California

ABSTRACT

Aqueous suspensions of ceramic powders are used in a variety of techniques for the fabrication of components, and the suspension properties control the resulting component microstructure and mechanical properties. The development and implementation of practical aqueous colloidal suspension characterization techniques using acoustophoresis is presented. The use of the acoustophoretic technique to optimize powder dispersion and compatibility so that desired suspension rheology, stability, and solids content are achieved is discussed. The example ceramic system discussed is the silicon nitride - yttrium oxide (added as a sintering aid) multicomponent system. Topics include technique development and implementation, suspension aging evaluation, effect of electrolyte additions, acid-base and surfactant titrations, effect of solids content, comparison with other suspension characterization techniques, correlation of colloidal dispersion properties with other powder properties such as oxygen content, and evaluation of multicomponent suspensions.

INTRODUCTION

Silicon nitride structural ceramic components are fabricated by liquid phase sintering of powder compacts composed of a mixture of the silicon nitride powder and sintering aid powders (typically rare earth oxides such as yttrium oxide). Generation of silicon nitride components with high strengths, reliability (Weibull modulus), and dimensional control places many requirements on the as-formed powder compact, such as a homogeneous distribution of Si_3N_4 and sintering aid powders with

no agglomerates and uniform high green density. If colloidal
filtration is the powder compact forming route, then the
suspension properties will control the resulting powder compact
properties. If it is assumed that the powder composition and
particle sizes/distributions are constrained by the final desired
microstructure and properties, then the colloidal suspension
properties that can be manipulated are the degree of dispersion
of the powders and the suspension solids content.

The dispersion of colloidal particles in an aqueous medium by the
generation of electrostatic or steric repulsion has been
extensively discussed in the literature [1-5]. In the case of
electrostatic forces, the generation of high zeta potentials is
generally observed to result in stable dispersions through the
generation of large mutually repulsive forces between particles.
The suspension will pack into high density compacts during
filtration due to the mutual repulsion between particles.
Low zeta potentials result in the domination of short distance
attractive forces such as Van der Waals forces, which result in
the generation of poorly packed floc structures (agglomerates)
which rapidly settle out and generate increased suspension
viscosities and instability. The agglomerates pack poorly,
resulting in inhomogeneous, low packing density compacts.

For multicomponent powder suspensions, powders with same-sign
large magnitude zeta potentials exhibit well dispersed stable
suspensions, while powders with opposite sign zeta potentials
exhibit hetero-coagulation, instability, and rapid settling [6].
Hetero-coagulation can be a benefit and has been utilized in
suspensions where the powder constituents have greatly different
densities and differential segregation can occur in a highly
dispersed suspension [7-9]. Electrostatic forces are typically
manipulated by changing pH of the suspension through the addition
of acids or bases.

Steric repulsion can be generated simultaneously with
electrostatic repulsion by the addition of long chain polyion
dispersants to the suspension. One end of the polymer attaches
to the particle surface, while the other end extends out and
repels like-polymer chain ends attached to other particles. If
the polymer is added to particles with low surface charge, the
charged polymers also generate a large charge and create an
appreciable electrostatic repulsion[10]. In practice, pH
manipulation and polyelectrolyte additions are used in
combination to stabilize structural ceramic suspensions.

Suspension solids content also influences a number of suspension
and cast part properties. Besides exhibiting higher viscosities,

           Characterization Techniques for the Solid-Solution Interface

higher solids content suspensions may exhibit greater tendencies
towards shear thinning and instability (manifested as gelation)
because of decreased particle repulsive forces due to double
layer overlap and compression when the particle separation
distances in the suspension approach the particle double layer
thicknesses [11,12]. As a result, pourability and conformability
of suspensions in molds will be reduced, as well as the time over
which the suspensions can be used. Higher solids contents do
result in shorter casting times and higher powder compact packing
densities, due to the smaller amount of liquid to be removed
during filtration and the smaller amount of rearrangement needed
for particles transitioning from the suspension to cast
compact[13].

Thus, the characterization of colloidal suspensions is critical
to allowing insight into suspension properties and how they can
be manipulated into generating the desired suspension
characteristics and resulting powder compact properties. Many
techniques are used to characterize colloidal suspensions.
Techniques that measure macroscopic suspension properties include
rheological measurements, settling tests, and measurement of
packing densities of cast powder compacts. But insight into the
fundamental nature of particle interactions allows the highest
degree of understanding of colloidal suspension stability, and
subsequently its manipulation. The most widely used technique
is electrophoresis, which allows determination of electophoretic
mobility as a function of suspension properties, such as pH and
electrolyte concentration. Electrophoresis is limited to very
dilute suspensions.

Recently acoustophoresis equipment has been developed that
utilizes high frequency electric fields to generate relative
motion in colloidal particles with surface charge in suspension
called the electrokinetic sonic amplitude (ESA) effect. The
movement of the particles generates a high frequency sound wave
which is related to the charge on the particle surfaces
and from which the electrophoretic mobility and zeta potential of
the particles can be calculated. The opposite procedure by which
high frequency sound applied to a suspension generates movement
of colloidal particles and results in an electric potential
related to the particle electrophoretic mobility, is called
ultrasonic vibration potential (UVP). The mobility of particles
in suspension can be monitored as the pH of the suspension is
varied, and as surfactant is added. The acoustophoresis
equipment is capable of measuring highly loaded colloidal
suspensions, up to 50 vol %.

The acoustophoresis technique has been evaluated, developed, and implemented for the characterization of silicon nitride based aqueous suspensions, to be used in the fabrication of silicon nitride structural ceramics by colloidal filtration. The following sections will discuss the development and implementation of acoustophoresis characterization, comparison and correlation with other colloidal suspension characterization techniques, examination of pH and surfactant titration techniques, effect of suspension solids content, and applications correlating colloidal properties with other powder characterisitics.

ACOUSTOPHORESIS TECHNIQUE DEVELOPMENT

Acoustophoresis techniques have been developed and refined at Garrett Ceramic Components over the last five years, principally for application with silicon nitride based aqueous suspensions. The first step in the characterization process is to prepare the colloidal suspension. For characterization of a single powder suspensions, a 1 vol % solids suspension is prepared in a teflon container (nominal suspension size is 250 ml). The powder is added to deionized water and the suspension is mixed for 15 minutes using a teflon coated magnetic stir bar. The stir bar speed is adjusted to ensure that no powder settles to the bottom of the container. The suspension is then ultrasonicated using an ultrasonic horn for 1 minute at 100 watts power, to break up any agglomerates and assure that maximum colloid surface area is available for evaluation.

The suspension is then placed into the acoustophoresis measurement equipment[a] and constantly stirred. The suspension properties are then monitored until they have equilibrated. Properties monitored are zeta potential, suspension pH, and suspension conductivity. Through experience and knowledge of the planned suspension applications, the suspension is considered to be equilibrated when none of the properties has varied more than 5% in 15 minutes. The desired suspension stability time is 24 to 48 hours (the total amount of time from initiation of suspension preparation to completion of colloidal filtration). It has been demonstrated that silicon nitride powders do react in water, due to the relative instability of surface amine groups and their replacement on the surface over time by silanol groups[14]. But it has been observed that the yttrium oxide powder is much more reactive in water (as demonstrated by the aging results in Figure 1) then representative silicon nitride powders. Yttrium oxide powder hydrolyzes readily in water and is generally

[a] Matec Model 8000

observed to become unstable in aqueous silicon nitride- yttrium
oxide suspensions long before the silicon nitride powders. After
the powder suspension is determined to be equilibrated, the pH of
the suspension is noted, as it is a characteristic of the powder
as a result of its desorbed and dissolved surface species.

The properties of the suspension are then characterized.
Acoustophoresis measurements are made using the ESA method. The
UVP method is not used, as it requires knowing the suspension
high frequency conductivity, which is not measured. The zeta
potential is typically determined by titrating the pH from the
equilibrated suspension pH to either the desired low or high pH
endpoints. Thus two experiments are required to generate a pH
vs. zeta potential curve. Titrations are only performed in one
direction for each experiment, since this approximates suspension
preparation for actual components, i.e. the suspension is only
adjusted to its desired final pH and there is no buildup of
undesirable counterions. The titrants used are 2.0N HNO_3 and
2.0N NH_4OH, since this acid and base are used in production
suspension pH manipulation, as they contain no detrimental
cations or anions. The relative strength of the titrants limits
the achievable pH range to 2.0 -10.5. A 25 second hold is
utilized between titrant additions to allow the suspension time
to equilibrate. Figure 2 shows a typical zeta potential vs. pH
curve for a silicon nitride powder.

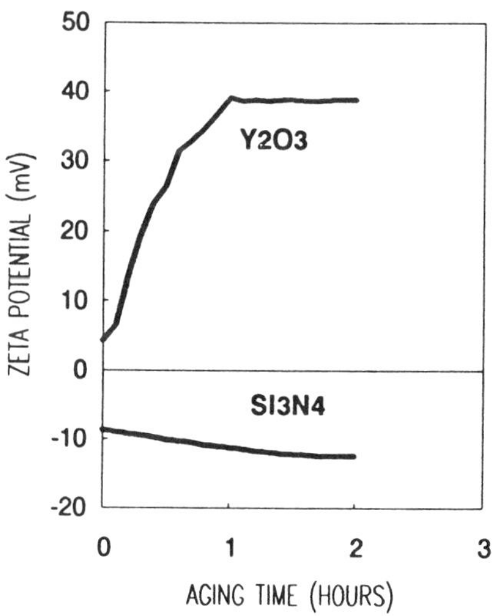

Figure 1. Aging Comparison of
Si_3N_4 and Y_2O_3 Powders in
Water.

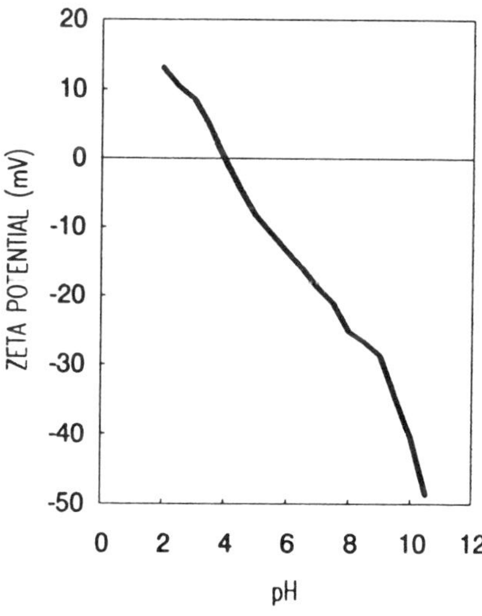

Figure 2. Typical Zeta Poten-
ial vs. pH Curve for Si_3N_4
Powder.

It is important to note that no electrolyte is added to these
suspensions. Electrolytes such as KCl and NaCl are typically
added to colloidal suspensions for electrophoresis experiments to
generate a constant ionic environment for suspension property
measurements. But alkali and alkaline earth cations and halide
anions are detrimental to silicon nitride (and many other
structural ceramics) high temperature properties and are not
added to colloidal suspensions used in component fabrication.
Experiments have been performed to evaluate the influence of
electrolyte additions on measured zeta potentials in pH
titrations of silicon nitride powders (see Figure 3). The only
influence of the electrolyte was to decrease the relative zeta
potential, with the greatest effect at low and high pH. Thus, no
detriment is observed by eliminating electrolyte additions for
acoustophoresis measurements.

The effect of surfactant additions on zeta potential is
determined by titrating the surfactants into suspensions. In some
cases, the pH is manually adjusted to be maintained at a desired
level, so that the effect of the surfactant can be determined at
a constant pH. The time between surfactant additions is 25
seconds, to allow for suspension equilibration after the previous
surfactant addition. A typical surfactant titration curve for
silicon nitride is presented in Figure 4.

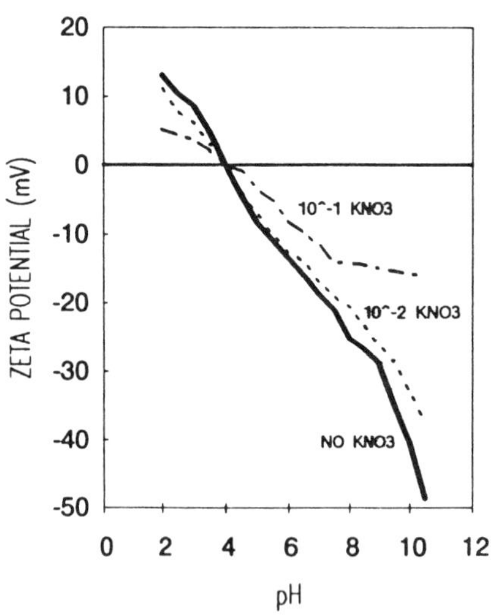

Figure 3. Effect of Electrolyte Additions on Zeta Potential Si_3N_4 Powder in Water.

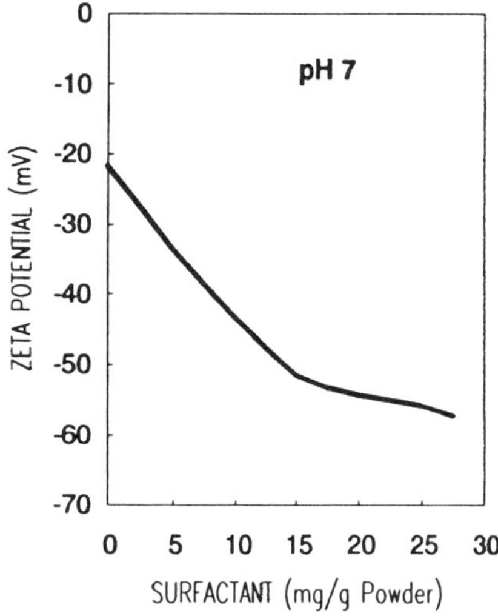

Figure 4. Example of Surfactant Titration Curve For Si_3N_4
Powder.

   Characterization Techniques for the Solid-Solution Interface

APPLICATION TO SILICON NITRIDE - YTTRIA SUSPENSIONS

Single Component Suspension Characterization

The first step taken in the development of silicon nitride based aqueous suspensions is to characterize the individual powder zeta potentials as a function of pH. In order to correlate zeta potential with other suspension characterization techniques, 5 vol% solids suspensions were utilized. Figure 5 shows the zeta potential vs. pH curves of silicon nitride (Denka SN-9S) and yttria (Shin-Etsu SU) powders. The Y_2O_3 was only evaluated above pH 7, since it dissolves under acidic conditions. The isoelectric point (iep) of the Si_3N_4 is 4.0 and the iep of the Y_2O_3 is 9.25. The powders exhibit opposite sign zeta potentials at pH 7. As the pH is increased, the Y_2O_3 moves through its iep and becomes negatively charged, the same sign charge as the Si_3N_4 powder. As pH 10.5 is reached, the two powders exhibit large zeta potential values.

The zeta potential values measured were correlated with three other suspension characterization methods - suspension viscosity, suspension sediment height, and suspension supernatent turbidity. The suspensions used for these methods were prepared at specific pH values, not titrated. The viscosity of the suspensions, measured using a high sensitivity couvette viscometer[b] is shown in Figure 6. Viscosity measurements correlate well with zeta potential values, i.e. viscosities are lowest where zeta potential values are highest, and the Y_2O_3 viscosity is highest at its iep, where there are no repulsive forces and floccing is allowed to occur.

Sediment height tests were conducted with equal amounts of suspension placed into test tubes for 48 hours. The relative sediment heights were then examined. The intrepretation of sediment height tests is that if a suspension is allowed to stand for an extended time, the particles will settle due to gravity. Powders that are well dispersed will pack into high density structures, forming a short sediment height. Poorly dispersed (low zeta potential) powders will pack poorly into low density structures, resulting in high sediment heights. Figure 7 shows relative sediment height results as a function of pH. The results correlate with zeta potential measurements in that highest sediment heights are achieved at the lowest zeta potential values.

Supernatent turbidity tests were conducted by examining suspensions placed in test tubes after 48 hours. The suspension

[b]Haake model CV-100

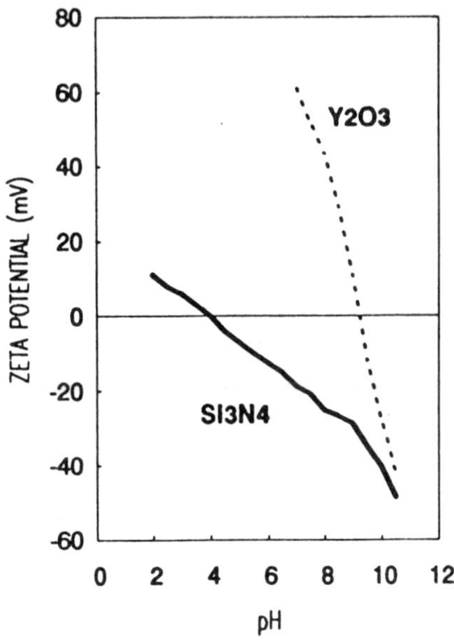

Figure 5. Zeta Potential vs. pH Curves for Si$_3$N$_4$ and Y$_2$O$_3$ Powders.

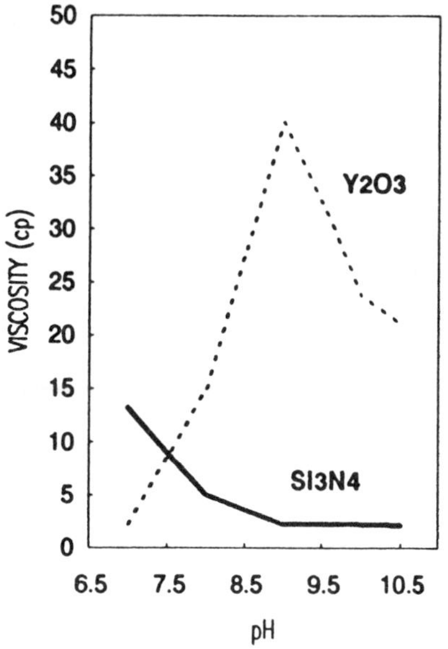

Figure 6. Viscosity of Si$_3$N$_4$ and Y$_2$O$_3$ Powders as a Function of pH.

supernatent turbidity was rated by visual inspection from 1 to 5, with 1 being clear and 5 being opaque. Interpretation of turbidity is that well dispersed particles will mutually repel one another and will stay suspended for long times. Poorly dispersed particles will floc and settle rapidly, leaving a clear supernatent. The results again correlate with zeta potential measurements in that opaque or nearly opaque supernatents are seen for high zeta potential particles, while a clear supernatent is observed for Y$_2$O$_3$ at its iep (see Figure 8).

The correlation is excellent for all three techniques, when correlated with the zeta potential measurement results, but acoustophoretic measurement of the zeta potential is both more rapid (two pH titrations vs. preparation of a number of different pH suspensions, and waiting for sedimentation to take place in the case of sediment height and supernatent turbidity tests), and it provides information on the fundamental forces controlling suspension properties.

Multicomponent Suspension Characterization
The second step taken in the development of Si$_3$N$_4$ - Y$_2$O$_3$ suspensions was to examine the effect of combining the two powders in suspension. Examination of the individual powder zeta

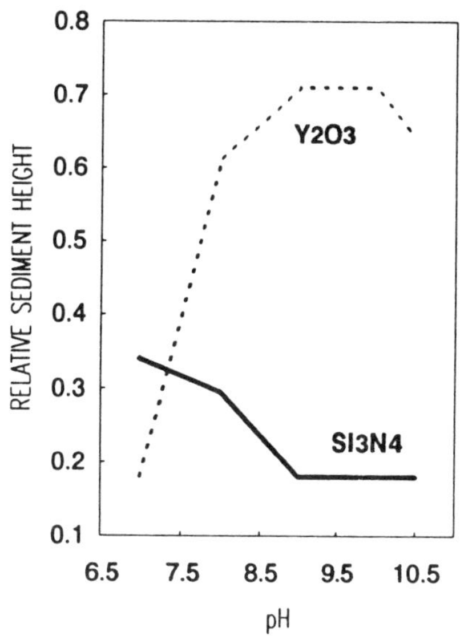

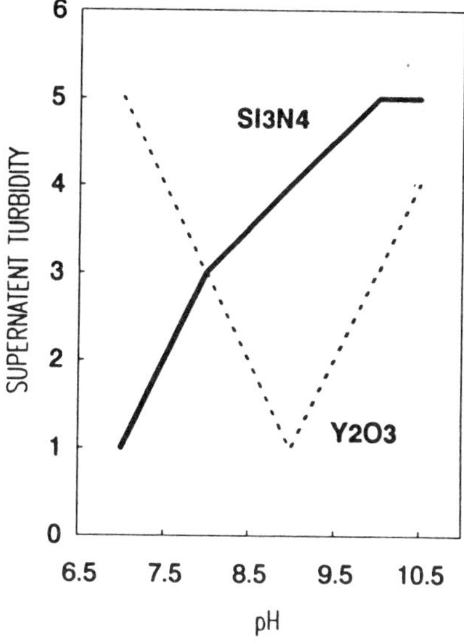

Figure 7. Sediment Height of Si_3N_4 and Y_2O_3 Suspensions as a Function of pH.

Figure 8. Supernatent Turbidity of Si_3N_4 and Y_2O_3 Suspensions as a Function of pH.

potential vs. pH curves in Figure 5 indicates that powders will have opposite sign zeta potentials at pH 7 and hetero-coagulation will occur. As the pH is increased, the Y_2O_3 will pass through its iep and become the same sign charge. Mutual repulsion will then occur with dispersion being maximized at pH 10.5, where both powders exhibit large magnitude zeta potentials. To determine if the individual zeta potential values of the two powders could be used to predict multicomponent suspension properties, two suspension compositions were prepared - 50 wt% Si_3N_4/50 wt% Y_2O_3, and 95 wt% Si_3N_4/5 wt% Y_2O_3. The suspensions were prepared at specific pH values of 7, 8, 9, 10, and 10.5. Earlier experiments have indicated that titrating the suspensions through regions of opposite sign zeta potentials induces hetero-coagulation that can not be reversed as the titration is continued. The results are presented in Figure 9. Zeta potential calculations used average values for the two powders based on their percent addition. For the 95 wt% Si_3N_4 suspension, the zeta potential measured is essentially that of the Si_3N_4 powder by itself in suspension. This can be expected due to the fact that the acoustophoresis equipment is measuring the resulting zeta potential of the entire suspension and that 5% Y_2O_3 will have a minimal effect on the suspension properties, even if it has an opposite sign zeta potential below pH 9, compared to the Si_3N_4. On the other hand,

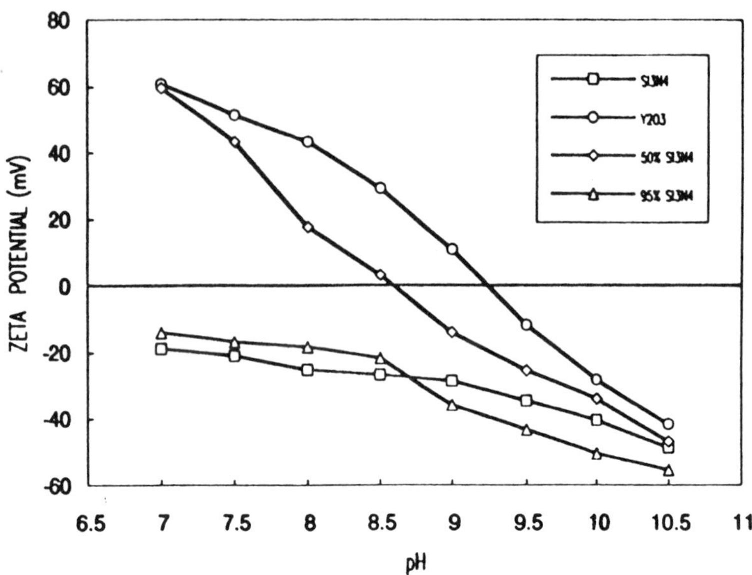

Figure 9. Comparison of Multicomponent Si_3N_4/Y_2O_3 Suspension Zeta Potential as a Function of pH, with the Separate Components.

the 50 wt% Si_3N_4 suspension shows a dramatic effect as a function of pH. Its measured zeta potential is approximately halfway between that of the Si_3N_4 and Y_2O_3 powders between pH values 8.8 and 10.5. This can be expected since the powders are both negatively charged and the suspension zeta potential should be approximately an average between the two powder zeta potentials. Below pH 8, the zeta potential tracks with the Y_2O_3 zeta potential, indicating that either the Y_2O_3 particles have hetero-coagulated onto the larger Si_3N_4 particles, thereby shielding the Si_3N_4 surface charge, or possibly that Yttrium cations have dissolved from the Y_2O_3 particle surfaces and adsorbed onto the Si_3N_4 surfaces, a process observed in other oxide materials by Wiese and Healy[14-15].

The viscosity of the two multicomponent suspensions was also measured and the results are presented in Figure 10. The 95 wt% Si_3N_4 suspension tracks with the viscosity of the Si_3N_4 suspension, while the 50 wt% Si_3N_4 suspension viscosity tracks with its measured zeta potential. At pH 7, although strong hetero-coagulation is predicted from the individual powder suspension zeta potentials, the result is a low viscosity suspension with high zeta potential. In addition, a sediment height evaluation was performed at pH 7 on the 59 wt% Si_3N_4

   Characterization Techniques for the Solid-Solution Interface

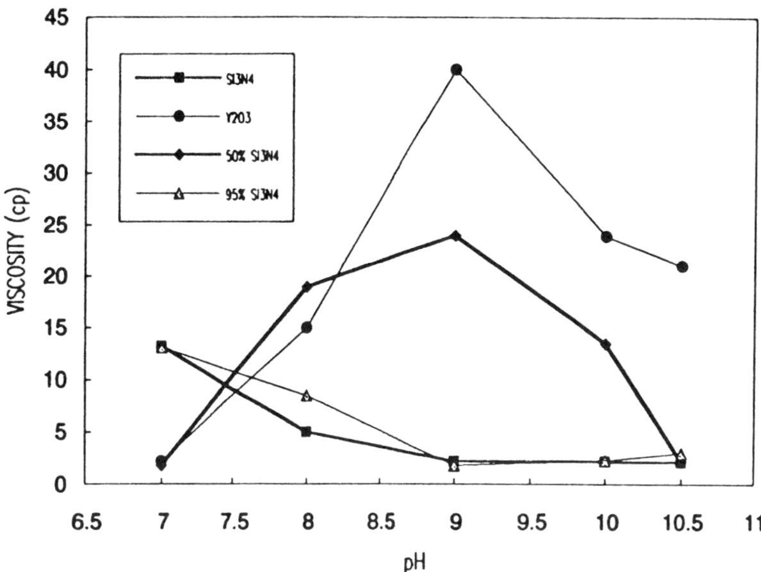

Figure 10. Viscosity of Multicomponent Suspensions as a Function of pH, Compared with the Separate Components.

suspension and it showed a higher sediment height than either of the individual powder suspensions at pH 7. This indicates that there is some type of hetero-coagulation occurring, but that it is not stongly influencing rheology.

In summary, evaluation of multicomponent suspensions by acoustophoresis is valuable in determining how the individual constituents will react when placed together, and how they can be manipulated, even though the acoustophoresis technique only measures the overall suspension zeta potential. But in order to effectively interpret results, the zeta potential of the individual powder should be known as a function of pH, and suspension characterization techniques such as viscosity measurements, can provide additional useful information.

Manipulation of Zeta Potential Sign by Dispersants
Multicomponent suspensions have a desired pH level at which they must be prepared, in order to maximize the zeta potential and degree of dispersion by the major powder component. Many times the minor powder additions have the opposite sign zeta potential of the major powder constituent at that pH and if added directly to the suspension will result in hetero-coagulation. The zeta potential sign can potentially be manipulated by addition of a

surfactant and the result can be measured using acoustophoresis. A specific example is a Si_3N_4-based suspension. The Si_3N_4 powder shown in Figure 2 can be most effectively dispersed at a pH range from 9 to 10.5. But a selected Y_2O_3 powder[c] has the opposite sign zeta potential (positive) at that pH range. In order to use the powder, the zeta potential was changed to negative by the addition of a sodium polyacrylate surfactant.[d] The surfactant titration curve is shown in Figure 11. Obviously, the Y_2O_3 must be predispersed with the surfactant before addition to the Si_3N_4, and the Si_3N_4 powder must also be compatible with the surfactant.

Effect of Solids Content

Acoustophoresis has the ability to measure zeta potential in high solids content suspensions. The equations relating the electro-acoustic effects to mobility and zeta potential are based on dilute suspensions of particles. As the solids content is increased to where electrical double-layer interactions and hydrodynamic interactions occur, the relationships become non-linear. At even higher solids contents, particle-particle interactions occur and reduce dynamic mobility. For most aqueous

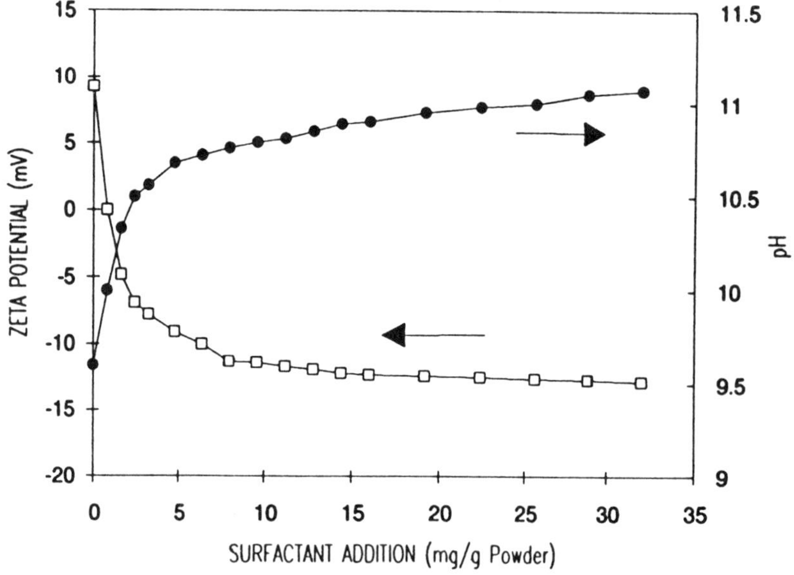

Figure 11. Reversal of Zeta Potential Sign of Y_2O_3 Powder in Aqueous Suspension by Surfactant Additions.

[c]Molycorp #5600 Y_2O_3
[d]W.R. Grace Daxad 30

suspensions, the non-linear behavior occurs above approximately
10 vol %. Si_3N_4 aqueous suspensions were prepared at a range of
solids contents to evaluate acoustophoretic zeta potential
measurements. The results are presented in Figure 12. The
calculated zeta potential is in fact the same for 10 vol % solids
content and less. The calculated zeta potential does decrease
above 10 vol % solids. Viscosity was measured for each
suspension prepared and is also shown in Figure 12. The
viscosity does not start to increase appreciably until 20 vol%
solids.

In order to determine the usefulness of acoustophoresis at high
solids loadings, Si_3N_4 suspensions and 95% Si_3N_4/5% Y_2O_3
suspensions were prepared at 37.7 vol% solids and at pH 9 and
10.5. The suspensions were characterized using acoustophoresis
and viscosity measurements. The results are presented in Figure
13. The comparison of 5 wt% Y_2O_3 substitution indicates
correlating measurement results - a zeta potential decrease and
viscosity increase, although the viscosity is much more sensitive

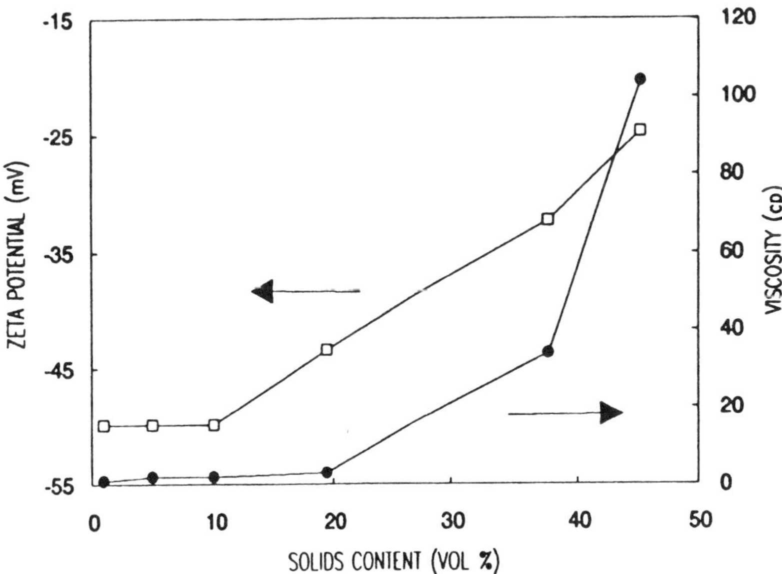

Figure 12. Effect of Si_3N_4 Suspension Solids Content on Zeta
Potential Measured by Electrokinetic Sonic Analysis.

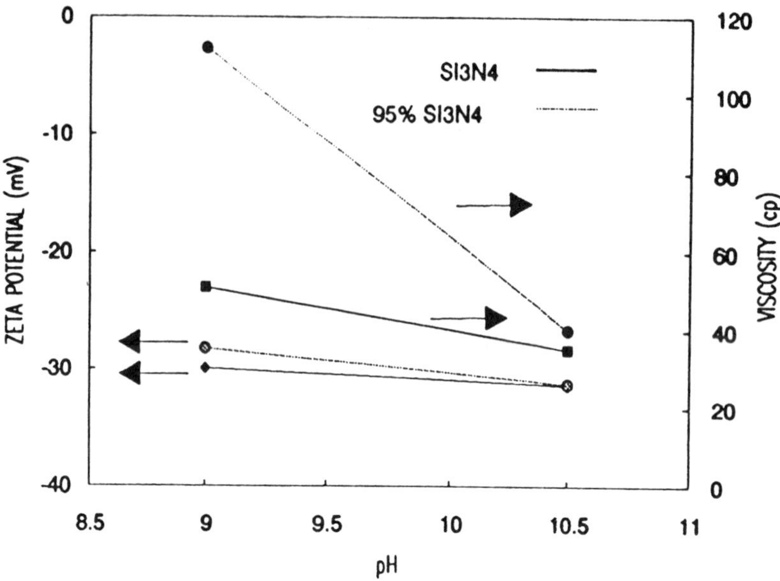

Figure 13. Comparison of Viscosity and Measured Zeta Potential of 37.7 vol% Solids Suspensions of Si_3N_4 and $95\%Si_3N_4/5\%$ Y_2O_3.

to small changes at the high solids content. Thus, to evaluate high solids content suspensions, viscosity measurements are a more desirable choice due to their greater sensitivity.

Correlation of Zeta Potential with Si_3N_4 Oxygen Content
Zeta potential measurements can also be used to evaluate other properties that affect zeta potential of powders. An example is the correlation that can be made between Si_3N_4 iep and oxygen content. In an experiment to evaluate this correlation, Si_3N_4 aqueous suspensions were milled in deionized water for a range of times, up to 168 hours. For each milling time, the suspensions were titrated using acoustophoresis to determine the isoelectric point, then the suspension was quickly dried and the oxygen content of the powders determined using a fusion technique.[e] The results are presented in Figure 14. The Si_3N_4 powder iep decreases in pH with increasing oxygen content. This is a result of the substitution of surface amine groups by silanol groups, which result in the surface becoming more SiO_2-like. The Si_3N_4 iep varies nearly linearly with oxygen content up to approximately 2.5 wt% oxygen, then the oxygen content increases rapidly with little change in iep. The transition at 2.5 wt%

[e]LECO Model TC-346 Oxygen/Nitrogen Analyzer

    Characterization Techniques for the Solid-Solution Interface

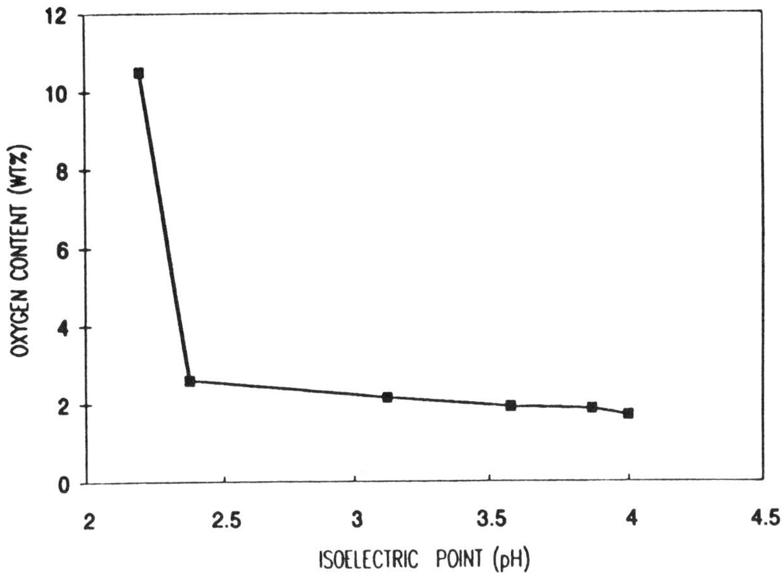

Figure 14. Correlation of the Isoelectric Point of Si_3N_4 in Aqueous Suspension (Milled at Various Times) with Oxygen Content.

oxygen potentially indicates the complete coverage of the Si_3N_4 particle surface by silanol groups. Further increases in oxygen content would not be expected to affect the Si_3N_4 iep, since the surface will continue to be fully covered with silanol groups.

SUMMARY

In conclusion, the development and implementation of acoustophoresis techniques for the determination of zeta potential in aqueous ceramic powder suspensions has been described. The acoustophoretic measurements correlate well with other suspension characterization methods for low solids content suspensions and are more rapid than conventional methods. At high solids contents, viscosity is a more sensitive indicator of suspension properties. Acoustophoresis techniques can also be used to characterize multicomponent powder suspensions, if single component suspension properties of the constituent powders are known. Powder properties which affect zeta potential, such as Si_3N_4 powder surface oxygen content, can also be correlated with zeta potential measured by acoustophoresis.

Characterization Techniques for the Solid-Solution Interface

REFERENCES

1. J.T.G. Overbeek, "Recent Developments in the Under-standing of Colloid Stability," J. Colloid Interface Sci., **58** [2] 408-22 (1977).

2. J.T.G. Overbeek, "How Colloid Stability Affects the Behavior of Suspensions," in _Emergent Process Methods for High Technology Ceramics_, edited by R.F. Davis, H. Palmour III, and R.L. Porter, Plenum Press, NY, 25-44 (1982).

3. J Lyklema, "Interfacial Electrochemistry of Disperse Systems," ibid 1-24. .

4. M.D. Sacks et. al., "Dispersion and Rheology in Ceramic Processing," in _Ceramic Powder Science_, edited by G.L. Messing, K.S. Mazdiyasni, J.W. McCauley, and R.A. Haber, Am. Ceram. Soc., Ohio, 495-516 (1987).

5. R.O. James, "Characterization of Colloids in Aqueous Systems," in _Ceramic Powder Science_, edited by G.L. Messing, K.S. Mazdiyasni, J.W. McCauley, and R.A. Haber, Am. Ceram. Soc., Ohio, 350-410 (1987).

6. T.W. Healy, G.R. Wiese, D.E. Yates, and B.V. Kavanaugh, "Hetero-coagulation in Mixed Oxide Colloidal Dispersions," J. Colloid Interface Sci., **42** [3] 647-9 (1973).

7. E. Carlstrom and F.F. Lange, "Mixing of Flocced Suspensions," J. Am. Ceram. Soc., **67** [8] C169-70 (1984).

8. F.F. Lange, B.I. Davis, and E. Wright, "Processing-Related Fracture Origins: IV, Elimination of Voids Produced by Organic Inclusions," J. Am. Ceram. Soc., **69** [1] 66-69 (1986).

9. T.M. Shaw and B.A. Pethica, "Preparation and Sintering of Homogeneous Silicon Nitride Green Compacts," J. Am. Ceram. Soc., **69** [2] 88-93 (1986).

10. R.G. Horn, "Surface Forces and Their Action in Ceramic Materials," J. Am. Ceram. Soc., **73** [5] 1117-35 (1990).

11. W. Albers and J.T.G. Overbeek, "Stability of Emulsions of Water in Oil: II. Charge as a Factor of Stabilization Against Flocculation," J. Colloid Sci., **14** [5] 510-18 (1959).

12. E.M. Deliso, A.S. Rao, and W.R. Cannon, "Electrokinetic Behavior of Al_2O_3 and ZrO_2 Powders in Dilute and Concentrated Aqueous Dispersions," in _Ceramic Powder Science_, edited by G.L. Messing, K.S. Mazdiyasni, J.W. McCauley, and R.A. Haber, Am. Ceram. Soc., Ohio, 525-536 (1987).

13. N. Ise, et. al., "Ordering of Charged Particles in Solution," Naturwissenschaften, **69**, S.544 (1982).

14. G.R. Wiese and T.W. Healy, "Coagulation and Electrokinetic Behavior of TiO_2 and Al_2O_3 Colloidal Dispersions," J. Colloid Interface Sci., **51** [3] 427-33 (1975).

15. G.R. Wiese and T.W. Healy, "Adsorption of Al(III) at the TiO_2-H_2O Interface," J. Colloid Interface Sci., **51** [3] 434-42 (1975).

     Characterization Techniques for the Solid-Solution Interface

ELECTROKINETIC SONIC ANALYSIS OF SILICON NITRIDE SUSPENSIONS

V. A. Hackley, R. S. Premachandran and S. G. Malghan

Ceramics Division
Materials Science and Engineering Laboratory
National Institute of Standards and Technology
Gaithersburg, MD 20899

In colloidal processing of ceramic slips, reactions occurring at the solid-solution interface play a dominant role in the dispersion properties. Electrokinetic sonic amplitude (ESA) measurements have been used to characterize the aqueous interfacial chemistry of Si_3N_4 suspensions. The ESA signal is proportional to the dynamic mobility and depends on the colloid shear-plane potential. Samples were subjected to acidic and alkaline soxhlet extraction, then titrated potentiometrically. Variations in the ESA signal and isoelectric pH indicate differences in the surface charge behavior of each powder. These differences may arise from surface contamination by specifically adsorbed ionic species or modification of potential determining surface sites as a result of treatment history. This paper is intended to provide hands-on information regarding the electrokinetic sonic analysis of aqueous ceramic dispersions.

INTRODUCTION

Colloidal processing of ceramic powders requires stringent control of particle dispersion in a liquid medium. Dispersion in aqueous solution is dependent largely on the surface electrochemical properties of the powders, and can be monitored using techniques that are sensitive to the shear-plane or zeta potential.[1] Ultrasonic techniques based on the electro-acoustic effect[2] have been used to measure electrokinetic properties of colloidal suspensions at solids concentrations where optical methods are incompatible.

This paper discusses the analysis of moderately concentrated aqueous silicon nitride suspensions using electrokinetic sonic amplitude (ESA) measurements. In this technique, an alternating electric field applied to a suspension causes the charged particles to oscillate. As a result of the density difference between the particles and medium, a sound wave (sonic field) is produced at the applied frequency (~ 1 MHz) with an amplitude proportional to the dynamic electrokinetic mobility. ESA measurements as a function of pH reveal the surface charge behavior of suspensions, and enable determination of the isoelectric point (IEP). The IEP is the pH of zero mobility and zeta potential, and is invaluable in characterizing surface chemistry and obtaining optimum processing conditions for a powder. A detailed description of the experimental procedures used in the preparation and analysis of powders is given, along with some limitations and potential difficulties associated with this measurement. Results are reported from a soxhlet extraction procedure for aqueous washing of Si_3N_4 powders.

EXPERIMENTAL PROCEDURE

Powders

Commercial Si_3N_4 submicron powders used in the present study are referred to in this paper as Powder A (SN E-10, UBE, Tokyo, Japan)[*], Powder B (LC 10, H. C. Starck, Berlin, Germany) and Powder C (Siconide P95M, KemaNord, Ljungaverk, Sweden). Physical properties of the as-received powders are listed in Table 1. Powder A is produced by reaction of

Table 1. Physical properties of Si_3N_4 powders.

Powder	A	B	C
Surface Area (m^2/g)[a]	10.7	12.6	6.1
Mean Diameter (μm)[b]	0.40	0.41	0.77
α/β Phase (%) [c]	95/5	95/5	91/9
Total O (wt%) [c]	1.25	1.90	0.99

[a] Quantachrome Autosorb-1 multi-point BET.
[b] Horiba LA500 optical diffraction.
[c] Manufacturer's specifications.

$SiCl_4$ and NH_3 to form silicon diimide, which is thermally decomposed to Si_3N_4. Powders B and C are both produced by nitridation of silicon powder, however Powder B is subsequently leached in HF to remove residual silicon.

As-received powders (AR) were washed by the method of soxhlet extraction in a standard pyrex arrangement. About 20 g of powder was placed in a Whatman filter thimble and extracted for 18 hrs with pH 2 HNO_3 (AW) or pH 10 NH_4OH (BW). Cleaned powders were dried at $50^{\circ}C$ under moderate vacuum and stored in plastic bags in a vacuum desiccator prior to analysis. Several batches of each powder were cleaned and combined to obtain a sufficient amount.

Sample Preparation

Suspensions (250 ml volume) for acoustic analysis were prepared at 1 volume % solids just prior to analysis. A weighed subsample of powder was added to the appropriate volume of deionized water and stirred for 10 min to facilitate wetting and equilibration. The suspension was dispersed by ultrasonic treatment with an immersed probe at 40 watts output power for three intermittent three minute periods. An ice bath was used to cool samples after each treatment so that suspension temperature did not exceed $35^{\circ}C$. Sample cooling is paramount, because the sample will not dissipate heat quickly and sharp increases in temperature will occur as the result of cavitation. Higher solids concentrations may require longer or more intense ultrasonic treatment to obtain sufficient dispersion.

Acoustic Analysis

Electroacoustic measurements and potentiometric titrations were performed using an Electrokinetic Sonic Analysis System from Matec Applied Sciences (Hopkinton, MA). Suspensions were placed in a teflon sample cell, and stirred with an overhead propeller and magnetic stir bar. In this batch mode, it is imperative that the dispersion remain suspended and well mixed throughout the duration of an experiment. This is particularly difficult during an acid-base titration when the IEP is crossed. Rapid coagulation at the IEP can lead to sudden changes in dispersion. Any settling or concentration gradient will adversely affect the ESA measurement, as the signal is proportional to the particle volume fraction ϕ in the neighborhood of the detector.

The phase angle between applied and measured fields must be calibrated for the sample after determining polarity against a known material (e.g., Ludox™ silica has negative polarity in water at any pH). Problems may be encountered when determining polarity of a suspension at a pH near the IEP, due to the low signal intensity. However, an incorrect choice for polarity will not affect the ESA measurement, only the apparent sign. In addition, the phase angle is used to determine the change in polarity of the suspension at the IEP. The most important system parameter is the frequency of the applied field. An autoscan of each sample will locate and lock on to the resonance frequency (strongest signal); however, for comparative purposes the frequency should be fixed for similar samples in a given experiment.

The results of an acid-base titration can depend on the details of how a titration was conducted and sample preparation. Fast ($\sim$ 30-120 sec between successive additions) and slow ($>$ 20 min increment) titrations may give a significantly different ESA curve. For this study, we chose a 120 sec delay between additions and measurement to allow time for mixing and fast-reaction equilibration. More concentrated suspensions may require longer equilibration times. Several strategies are possible for obtaining ESA data over a broad pH range. (1) The starting pH may be preadjusted to one end of the analysis range prior to beginning a titration. (2) A sample may be titrated to one endpoint then backtitrated to the other extreme. (3) Separate acid and base titrations starting from the "natural" pH may be combined. The choice depends on the stability of the samples and whether or not preadjustment changes the surface chemistry. This must be established experimentally for each powder. The method of backtitrating, for example, often results in hysteresis, complicating interpretation of results. In the present study, titrations were performed at the natural pH and, if necessary, data from two separate runs were combined.

Titrants and supporting electrolytes should consist of ionic species that are chemically inert in the system to be studied (except for potential determining hydroxide and proton). Specifically adsorbed species will shift the IEP, and are therefore readily identifiable. Nitric acid and ammonium hydroxide were the titrants in the present study. No supporting electrolyte was used, except in a separate analysis of ionic effects involving Powder C.

RESULTS AND DISCUSSION

For a colloidal suspension, ESA is the measured pressure amplitude of the sonic wave per unit applied electric field (P/E), and will depend on several parameters.[2,3]

$$ESA(\omega) = \frac{P}{E} = c\,\Delta\rho\,\phi\,f_G\,\mu_d(\omega) \qquad (1)$$

Where ω is frequency, c is sound velocity, $\Delta\rho$ is the density difference between particles and medium, ϕ is the particle volume fraction and f_G is a geometrical factor that depends on electrode geometry. Dynamic mobility $\mu_d(\omega)$ is related to electrophoretic mobility μ_e through a frequency/size

    Characterization Techniques for the Solid-Solution Interface

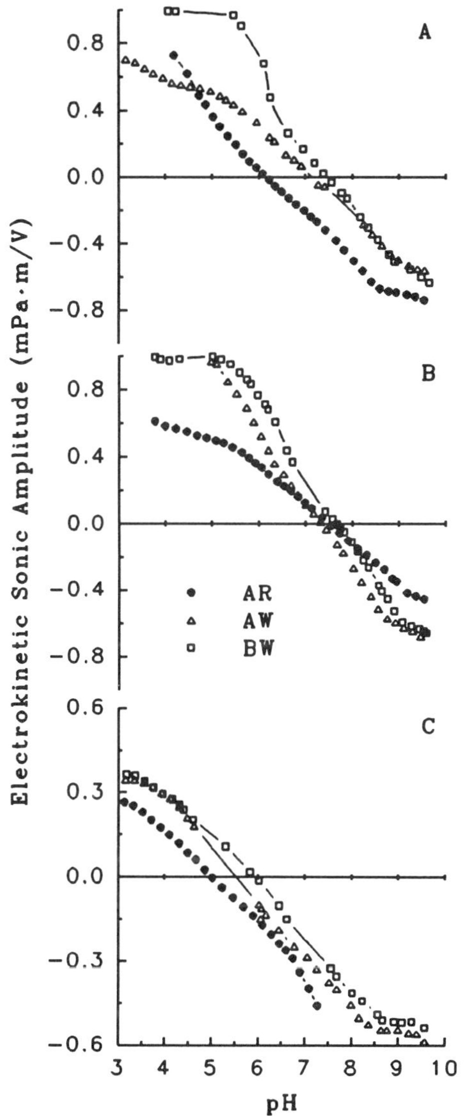

Figure 1. ESA vs. pH for Si_3N_4 Powders A, B and C: as-received (AR), acid (AW) and base (BW) washed.

dependent inertial term. Under certain conditions (i.e., moderately concentrated submicron particles with thin double-layers) $\mu_d(\omega)$ is a good approximation for μ_e.[4] The linear dependence of ESA on ϕ holds for particle concentrations up to 5-10% v/v, depending on the particular system. At higher concentrations, nonlinear effects from hydrodynamic and double-layer interactions become apparent.

Figure 1 shows ESA as a function of pH and pretreatment for powders A, B and C. The IEP is clearly defined in all curves, yielding values for the as-received powder of (A) 6.2 (B) 7.5 and (C) 5.0. These results compare well with reported electrophoresis measurements on dilute suspensions[5] which yielded 6.2, 6.5 and 4.2, respectively, for the same powders. In addition, the response to soxhlet extraction differs significantly between the three powders (Fig. 2). Powders A and C exhibit the greatest change in IEP, increasing about one pH unit after acid or base extraction; whereas, Powder B showed no significant shift in IEP as a result of extraction. In Figure 2 one can see a correlation between the conductivity/pH of sample suspensions (prior to analysis) and the corresponding IEP values. Powder B, with the most basic IEP, produced a much higher conductivity and a more acidic pH than the other two powders. This indicates leaching of an acidic species from the surface layer of Powder B. Indeed, Bergström and Pugh[5] found that fluoride is the dominant leachate from this powder, and may be attributed to the HF treatment during manufacture. After extraction, all suspension conductivities are similarly low ($\sim 10\mu$S/cm) and the powders exhibit a more basic IEP approaching that of Powder B.

Silicates (either dissolved or colloidal) and ammonia may also be released into solution.[5] These species are formed during hydrolysis of the Si_3N_4 surface:

$$6H_2O + Si_3N_4 \rightarrow 3SiO_2 + 4NH_3$$

To explain the present results, the following scenario is proposed. As the surface layer becomes oxygen rich through oxidation and hydrolysis reactions, the IEP is lowered toward that of pure SiO_2 (IEP ~ 2). Extraction with either acid or base solution removes silica from the surface and increases the IEP; base solution being slightly more effective. Greil and coworkers[6] found a linear relationship between the IEP and surface oxygen concentration for a series of Si_3N_4 powders. Following this line of thought, we may speculate Powder B does not respond to extraction because it contains low surface oxygen as a consequence of the previous HF treatment. Ammonia, which appears to be mostly physically adsorbed, is readily removed as a gas in the dry phase or in solution without greatly affecting surface chemistry.[7] Titrations in the presence of NH_4^+ salts do not indicate the shift in IEP that would suggest specific interactions with the Si_3N_4 surface.

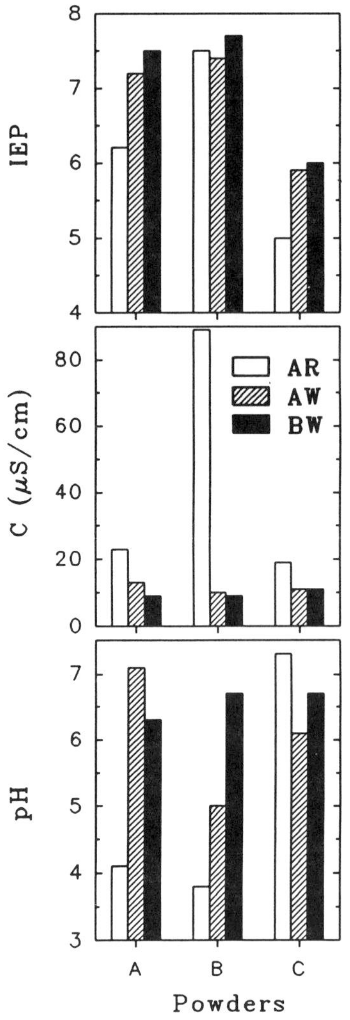

Figure 2. Suspension IEP, conductivity and pH for Si_3N_4 powders A, B and C: as-received (AR), acid (AW) and base (BW) washed.

   Characterization Techniques for the Solid-Solution Interface

When performing ESA measurements on suspensions in the presence of a supporting electrolyte, three separate effects must be considered. The first is screening of the electrically charged double-layer, which depresses the ESA signal but does not affect the IEP. The second effect occurs when one of the electrolyte ions interacts chemically with the particle surface. Specific interactions of this type are indicated by a shift in the IEP toward a more basic pH (cation adsorption) or more acidic pH (anion adsorption). The third effect, discussed below, is due to the electroacoustic properties of the ions themselves, and is independent of the powder.

Ionic contributions to the measured sonic amplitude are generally small relative to the particle signal, and can be swamped out by increasing the particle concentration sufficiently. The ionic amplitude will depend on the ion pairs and somewhat on the pH. Since cations tend to hydrolyze more than anions, their effective mass is usually greater. This imbalance leads to a net positive sonic amplitude added to that of the sample. An exception is NH_4NO_3, which gives a negative amplitude. Near the IEP this background signal can become significant enough to affect determination of the IEP and lead to erroneous results. Figure 3 illustrates the effect of increasing NaCl concentration on the ESA curve as a function of pH for Powder C. These results might lead to the false conclusion that Na^+ ions are specifically adsorbed on the Si_3N_4 surface. However, after subtracting the background amplitude measured separately for each electrolyte solution, the curves collapse to yield a similar IEP near pH 5.

The discontinuity near the apparant IEP in Figure 3 (see for example 0.05M NaCl) is an artifact, and is due to the manner in which the polarity is determined using the phase angle between the applied and measured fields. This discontinuity is symptomatic of an ionic background interference, and its

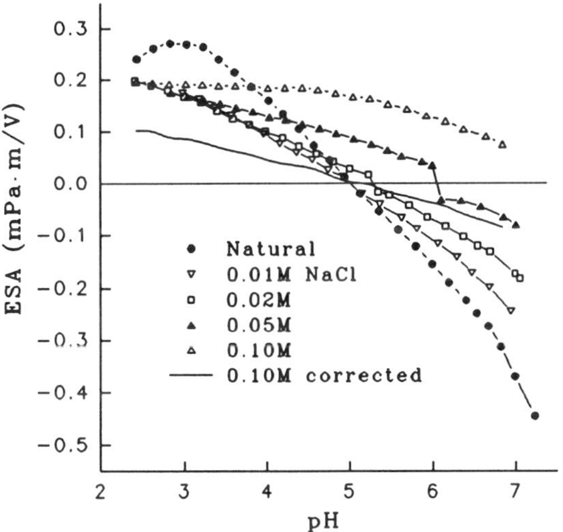

Figure 3. Effect of supporting electrolyte on ESA-pH curves for Si_3N_4 Powder C.

presence should warrant closer scrutiny of experimental results. Similarly, dependence of the IEP on particle volume fraction in the presence of a supporting electrolyte should be tested for background interference; an ionic contribution to the ESA would increase relative to the particle signal as the solids concentration decreased. Measurements of the pH dependence of the electrolyte sonic amplitude reveal a nearly constant magnitude above pH 5, and a monotonically decreasing function below this value. These effects are probably due to pH dependent changes in hydrolysis of the cations.

CONCLUSION

The aqueous surface chemistry of three silicon nitride powders was studied using electrokinetic sonic amplitude measurements. Differences in the isoelectric point (IEP) between the three powders are interpreted as arising from differences in the powder processing and subsequent treatment which affect the surface oxygen concentration. Powder surface cleaning was achieved by Soxhlet extraction. The magnitude of response to this cleaning procedure depended on the initial surface state (i.e., oxygen content) of each powder. Generally, the IEP tends to shift toward alkaline pH upon extraction, suggesting silica is removed from a hydrolyzed surface layer. The effects of supporting electrolytes on the ESA measurement are also discussed.

REFERENCES

1. R.J. Hunter, *Zeta Potential in Colloid Science*, Academic Press, New York, 1981.
2. R.W. O'Brien, "Electro-acoustic effects in a dilute suspension of spherical particles", *J. Fluid Mech.*, **190**, 71-86 (1988).
3. R.W. O'Brien, "The electroacoustic equations for a colloidal suspension", *J. Fluid Mech.*, **212**, 81-93 (1990).
4. R.W. O'Brien, B.R. Midmore, A. Lamb and R.J. Hunter, "Electroacoustic studies of moderately concentrated colloidal suspensions", *Disc. Faraday Soc.*, **90**, 301-12 (1990).
5. L. Bergström and R.J. Pugh, "Interfacial characterization of silicon nitride powders", *J. Am. Ceram. Soc.*, **72**, 103-09 (1989).
6. P. Greil, "Evaluation of oxygen content on silicon nitride powder surface from the measurement of the isoelectric point", *J. Euro. Ceram. Soc.*, **7**, 353-59 (1991).
7. P.K. Whitman and D.L. Feke, "Comparison of the surface charge behavior of commercial silicon nitride and silicon carbide powders", *J. Am. Ceram. Soc.*, **71**, 1086-93 (1988).

Spectroscopic Techniques

FOURIER TRANSFORM INFRARED SPECTROSCOPY
OF HYDROUS OXIDE/AQUEOUS SOLUTION INTERFACES

L.D. Tickanen*, M.I. Tejedor-Tejedor and M. A. Anderson
Water Chemistry Program
University of Wisconsin
660 North Park Street
Madison, Wisconsin 53706

ABSTRACT

In our program, we study the role of surface chemistry in the processing of ceramic powders such as alumina and silicon nitride. In addition, we are interested in the surface chemistry of metal oxide particles, both in suspensions and when sintered to form ceramic membranes. In this paper, we review several applications of attenuated total reflection-Fourier transform infrared (ATR-FTIR) spectroscopy for *in-situ* surface chemical studies of metal oxide particles. For qualitative and quantitative studies of aqueous suspensions of metal oxides, we demonstrate the application of one specialized form of ATR-FTIR spectroscopy, namely, cylindrical internal reflection (CIR)-FTIR spectroscopy. In so doing, we discuss several factors and experimental conditions which can affect the qualitative and quantitative analyses of these suspensions. Finally, we illustrate the use of *in-situ* attenuated total reflection (ATR) cells which utilize "flat" optics in the study the hydrous oxide/aqueous solution interfaces present in particulate ceramic membranes. Because the membranes consist of a relatively thick film of sintered oxide particles, quantitative investigations of adsorption phenomena and the nature of "interfacial" water are facilitated.

INTRODUCTION

Metal oxides are important in the natural environment and in industrial applications. Most metal oxides have amphoteric surfaces that are capable of adsorbing molecules of many types, cations and anions, both inorganic and organic. In nature, this adsorbing capability serves as an important control on the distribution and aqueous solubility of trace metallic and nonmetallic ions, and various organic species. In industry, adsorption by metal oxides plays important roles in such areas as wastewater treatment, ion exchange, catalysis, and ceramic membrane separations. Thus, analytical techniques which allow the study of surface phenomena at the molecular level can allow us to better understand these processes.

Infrared (IR) spectroscopy is used to study molecular vibrations which result in a change in the electric dipole moment of the molecule. It has been used to study chemical bonds, to discern reaction mechanisms, and to describe the molecular motions in liquids. In IR spectroscopy, a sample is irradiated with radiation of wavelength in the range of 1-1000 microns. Sources are heating elements such as the globar. The infrared spectrum may be scanned using a monochrometer (dispersive IR) or a time domain spectrum (interferogram) may be collected and converted to a frequency domain spectrum using the Fourier Transform (FTIR). FTIR spectroscopy is well suited to the study of heterogeneous systems such as suspensions because several scans (interferograms) can be collected in a short period of time and coadded, thus greatly enhancing the signal-to-noise ratio. The short scanning times can allow one to obtain several short duration snapshots, during kinetics studies, for example.

More recently, IR spectroscopic techniques which rely upon the principle of total internal reflection have been used to study reactions at electrodes [1], the adsorption of solutes to colloidal particles [2,3], and the surfaces of thin films [4]. Such techniques are therefore well suited for the study of ceramic membranes and the oxide precursors from which they are produced. In our program, we have found FTIR spectroscopy to be a great asset in studying many of the phenomena involved in both the genesis and use of ceramic particles and membranes. We have been able to monitor steps that occur during the formation of primary particles and during the calcination process. We have also been able to study the behavior of adsorbates and interfacial water, in both qualitatively and quantitatively. In this paper, we hope to provide an overview of how FTIR spectroscopy can be an asset to both colloid chemists and ceramists by discussing applications to two general types of samples--aqueous suspensions and supported ceramic membranes. While we will not discuss detailed experimental procedures, as they are explained in many of the references cited, we will point out some of the interesting aspects and potential problems one might encounter while studying phenomena occurring in the interfacial regions surrounding metal oxide particles.

BACKGROUND THEORY

Over the past decade, attenuated total reflection (ATR) Fourier transform infrared (FTIR) spectroscopy has evolved into a valuable technique for the qualitative and quantitative investigation of aqueous suspensions and for studies of adsorption phenomena in aqueous systems. Suspensions consist of multiple phases, usually a bulk liquid phase surrounding particles that are often larger in diameter or length than the wavelength of visible light. Furthermore, the particles may be charged under a variety of conditions (pH and ionic strength), and can interact with the instrumentation that is being used to study them, through electrostatic or van der Waals' forces. For these reasons, suspensions present

     Characterization Techniques for the Solid-Solution Interface

unique challenges not often encountered in the analysis of more conventional systems (i.e. solids, liquids, or solutions). Studies of suspensions using techniques employing visible radiation are not very useful, with the exception of light scattering studies, and these studies are generally applicable to dilute suspensions.

On the other hand, many suspensions contain particles having sizes of the order of a micron or less, which enables their study using IR spectroscopy. Transmission IR spectroscopy is limited in usefulness by the strong absorption bands of water in the infrared. For cell thicknesses greater than about 10 microns, the bands of water attributed to O-H stretching and H-O-H bending transitions can obliterate most other features in the spectral regions of interest. On the other hand, path lengths of less than 10 microns would require the suspension sample to be confined in spaces so small that aggregation could be induced, changing the nature of the sample. Furthermore, scattering may still be a problem in transmission experiments if large particles are present.

ATR-FTIR spectroscopy can overcome many of these problems. This is because ATR spectroscopy is based upon the principle of total internal reflection, where the sample probe is an exponentially decaying evanescent wave which extends externally from the internal reflection element (IRE) into the sample. Thus no confinement of the sample is required, because the probe does not pass through an enclosed cell, but rather emanates outward from the IRE.

The sampling depth (d_s) in ATR spectroscopy, defined by Harrick [5], is the distance from the IRE at which the evanescent field is 5% of E_o, its value at the IRE surface ($E_o e^{-3}$):

$$d_s = \frac{3\lambda_1}{2\pi(\sin^2\theta - n_{21}^2)} \tag{1}$$

where λ_1 is the wavelength of the IR radiation *inside* the IRE, Θ is the angle of incidence, and n_{21} is the ratio of the refractive index of the sample to that of the IRE. Although strictly a definition, d_s allows one establish some appreciation as to the dimensions of the region being probed in the experiment. In the IR region of the spectrum, this distance is usually a few microns or less, which greatly reduces the dominance of the water bands in the spectrum. Figure 1 shows a schematic diagram of one type of ATR optics, as well as a representation of the evanescent field amplitude as it decays outward from the IRE.

In ATR studies, the IRE is generally chosen with properties which optimize the conditions for study of the sample (e.g. sampling depth). IREs can be fabricated from many materials which vary in such important properties as index of refraction, hardness, brittleness, aqueous solubility, and optical transparency. Selection of an appropriate IRE for a given application involves a number of other considerations, such as the critical angle when used to study a given sample, the depth of penetration of the evanescent wave into the sample, the range of the infrared spectrum that one wishes to observe, the spectrometer optics, chemical compatibility with the sample, and cost.

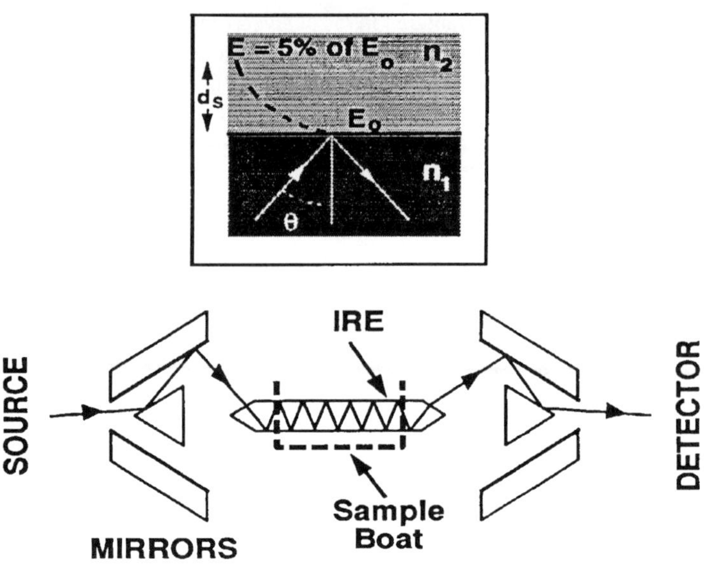

Figure 1. Diagram of an Cylindrical Internal Reflection (CIR) ATR cell, showing the path of the totally internally reflected IR beam. The sampling depth, which is given by equation (1), is shown in terms of the decaying evanescent field.

For liquids and other types of homogeneous samples, we can assume that the spectrum obtained for the sample being probed is independent of the sampling depth. However, this is not always true for suspensions and other heterogeneous systems. We have found, however, that the distribution of the suspension particles around the IRE usually can be described by one of three mathematical models which describe the distribution conceptually in terms of films of varying thickness relative to the sampling depth: thick, thin or intermediate. It is noteworthy to mention that the term "film", as used here, does not necessarily mean a uniform solid, but may imply a layer of any phase whose index of refraction is constant over a scale of homogeneity of the order of a wavelength of the probe radiation (several microns in the case of IR radiation). Thus, a film could be a bulk liquid, a ceramic membrane, or a uniform distribution of colloidal particles.

The two most important model cases (thick films and thin films) have been discussed by Harrick [5,6], where appropriately selected limits on the thickness give rise to results which can simplify the mathematics involved, leading to straightforward calculations. When the thickness of the absorbing medium is considerably greater than the sampling

   Characterization Techniques for the Solid-Solution Interface

depth (i.e. where the amplitude of the evanescent wave has decayed to a value small enough to be negligible), the absorbance is given by the expression for thick films (assuming unpolarized radiation:

$$a = \frac{N\alpha c\lambda_1 n_{21}\cos\Theta[(3 + n_{21}^2)\sin^2\Theta - 2n_{21}^2]}{2\pi(1 - n_{21}^2)[(1 + n_{21}^2)\sin^2\Theta - n_{21}^2](\sin^2\Theta - n_{21}^2)} \tag{2}$$

where α is the molar absorptivity of the species being studied, **c** is the concentration, and **N** is the number of reflections made by the IRE beam inside the IRE. All other definitions are the same as in equation (1).

Thus, the characteristics of ATR spectra, such as absorption intensities and absorbance band ratios, are determined not only by the concentration of species being investigated, but also by the index of refraction, and the thickness and the homogeneity of the study medium. For the situation where the thickness of the layer approaches zero (i.e. the sample film is so thin that all molecules in the film experience essentially the same interaction with the evanescent wave) we can derive a similar expression, although in this case, the absorbance is dependent not upon the wavelength, but upon the film thickness:

$$a = \frac{2N\alpha c n_{21}d\cos\Theta\ [(2+n_{31}^2+n_{32}^4)\sin^2\Theta-n_{31}^2]}{(1-n_{31}^2)\ [(1+n_{31}^2)\sin^2\Theta-n_{31}^2]} \tag{3}$$

where **d** is the actual thickness of the thin layer, n_{31} is the ratio of the index of refraction of the medium at thicknesses greater than **d** (i. e. the "external" medium) to that of the IRE, and n_{32} is the ratio of the index of refraction of the external medium to that of the film. Other definitions remain the same, as above. In this case, the absorbance is independent of the wavelength of probe radiation, so profiles of spectra obtained from thin films resemble transmission spectra. (For practical purposes, a film of thickness less than d_s /60 can be considered as a thin film).

It is important to emphasize that the spectral profiles obtained for films of different thickness are thus different. This fact is important in recognizing which type of system one is working with, and is crucial in allowing one to select the appropriate mathematical expressions for use in quantitative characterizations, which we will discuss later.

STUDIES OF COLLOIDAL SUSPENSIONS

In our earliest efforts, we demonstrated how ATR-FTIR spectroscopy can be routinely used to study the bonding structures of organic and inorganic ligands with the surface of a variety of metal oxide particles suspended in aqueous media [7-12]. In these studies, we found that a special application of ATR-FTIR spectroscopy, namely, cylindrical internal reflection (CIR)-FTIR spectroscopy, was most advantageous for studying suspensions.

This is mainly because the use of the CIR CIRCLE sample cells (Spectra Tech, Inc.) does not require that the sample be confined in small spaces, and allows for easy stirring of the sample. Furthermore, the CIR optics yield complete scrambling of the polarizations present in the IR beam, which simplifies the mathematics needed for quantitative work.

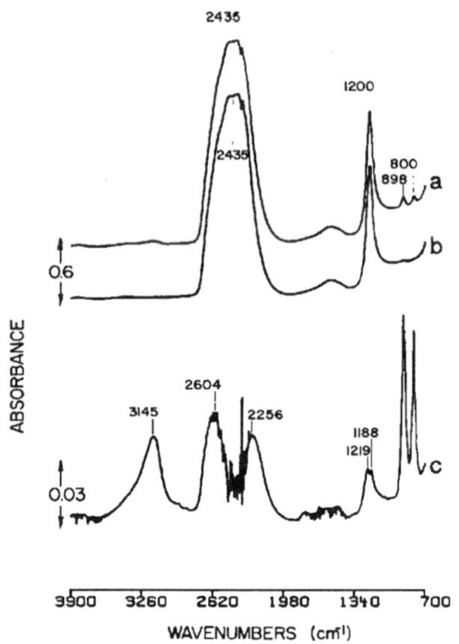

Figure 2. **CIR-IR spectra of a goethite/D_2O suspension (pD=7, solids concentration = 100 g/l): (a) suspension; (b)supernatant; (c) difference = (a)-(b).**

We developed a qualitative methodology for isolating the absorbance spectra of suspended particles and the associated interfacial material (water and adsorbed species) from the bulk solution (supernatant) in aqueous suspensions, which is illustrated in Figure 1. Such difference spectra resulted from the direct, simple subtraction of the spectrum of the supernatant liquid for each sample from the spectrum of the suspension (particles plus bulk liquid), with allowances being made for the volume occupied by the suspended particles in concentrated suspensions. Figure 2 shows the CIR-IR spectra of a suspension of colloidal goethite (α-FeOOH) particles in D_2O solution, of the supernatant, and of the goethite particles and the associated interfacial material (the difference spectrum). In the difference spectrum, absorption bands at 3145, 898, and 800 cm^{-1} are due to goethite vibrations. All other bands are related to the interfacial region. (We found it necessary to use deuterium oxide (D_2O) rather than plain water as the liquid phase in many cases, because the O-H stretching bands for water are so strong that they can hide the O-H stretching bands for the metal oxide particles. The O-D stretching frequency in D_2O is far removed from the positions of these metal oxide O-H bands).

We soon came to realize that, unlike in the case of dilute solutions, the intensity of the

                    Characterization Techniques for the Solid-Solution Interface

absorption bands for colloids was not only a function of the formal particle concentration but as well dependent also on properties of the suspension as pH, the ionic strength, the nature and concentration of dissolved species and whether the sample had been dispersed using sonication. From the way that these factors influence the band intensities it is clear that changes in the aggregation state of the particles are an important source of variability in the spectral bands. The intensity is generally found to decrease with increasing aggregation. Figure 3 depicts changes in spectra of α-FeOOH particles suspended in D_2O under various pD and ionic strength conditions. We also noted that the absorption band ratios for the various α-FeOOH bands *within a given spectrum* also suggested that the distribution of goethite particles within the sampling depth of the IRE was different for different suspension conditions. By changing the particle concentration, the magnitude and sign of the charge on the particles, as well as the ionic strength, we found that we could vary the thickness of the layer of goethite responsible for the spectral bands.

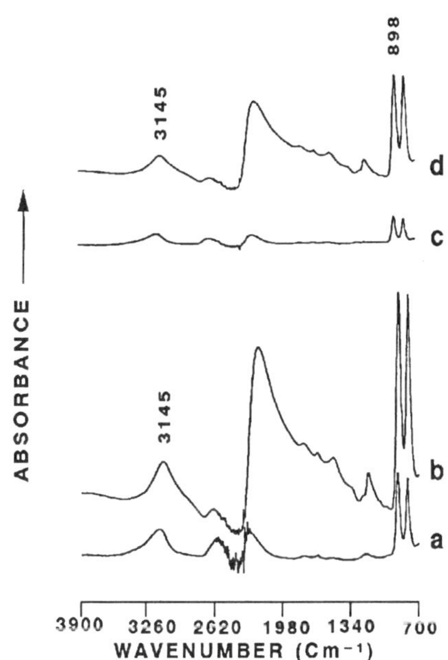

Figure 3. CIR-IR spectra of goethite in D_2O for different pD and ionic strength values: (a) pD=4, I=0.1 $\underline{M}$ NaCl; (b) and (d) pD=4, I=0.01 $\underline{M}$ NaCl; (c) pD=8, I=0.01 $\underline{M}$ NaCl.

Figure 4 shows one example of the variation of band ratios for goethite under different suspension conditions. The ratios for absorbance values at 900 cm⁻¹ versus 3150 cm⁻¹ range from the thin film limit of 1.5 (defined by the transmission spectrum **a**) for which $d < d_s/60$, to the thick film limit of 5.3 (product of the band ratio value for the thin film

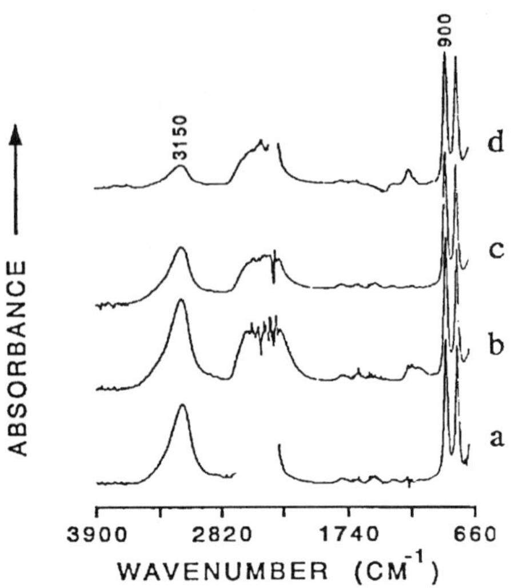

Figure 4. IR spectra of goethite particles: (a) transmission spectrum in KBr; (b) CIR spectrum in D_2O, pD=4 and solids concentration = 0.02 g/l; (c) CIR spectrum in D_2O, pD=8 and solids concentration = 40 g/l; (d) CIR spectrum in D_2O, pD=4 and solids concentration = 60 g/l.

multiplied by the ratio of the wavelengths at which the maxima occur) in which $d>d_s$, as is shown in spectrum **d**. It is assumed in performing the direct subtraction of the supernatant spectrum from the suspension spectrum, that the refractive indices of the suspension and the supernatant are the same over the range of frequencies studied--an assumption which may be erroneous in many cases. This fact is critical in the development of ATR-FTIR techniques for the quantitative investigation of suspensions and, later, ceramic films.

Complexation studies

Because ATR spectra of suspensions may contain contributions due to aggregation, particle distribution, and changes in refractive index, the ATR difference spectra produced by the simple subtraction of supernatant from suspension could not be readily used to measure concentration of either colloidal particles or interfacial species. Nonetheless, in these difference spectra, both the shape and the frequency at the maxima of spectral bands for interfacial species correspond to the true nature of the chemical structure of these species, except if they occur in the vicinity of strong absorption bands for the solvent. Consequently, they have provided valuable information on the nature of the bonds between interfacial species and the surface atoms of the colloidal particles. Using this methodology, we have been able to establish, for example the type of surface complexes that benzoic and phenolic compounds as well as some inorganic oxo-anions form with the surface of α-FeOOH [8,9] and anatase (TiO_2) [10].

In Figure 5, spectra **a** and **b** illustrate the differences between the spectrum of salicylate ion in D_2O solution (ionic salicylate) and the one of salicylate at the α-FeOOH/ D_2O solution interface, where salicylate forms a chelate complex with iron atoms at the surface he goethite particles. Figures **c** and **d** allow the comparison of vibrations of nitrate ion

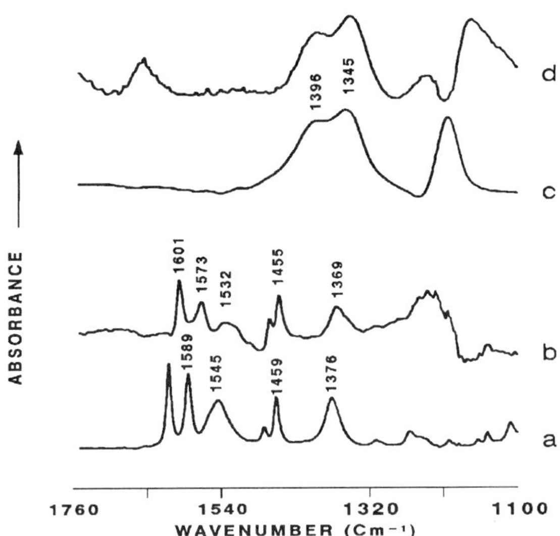

Figure 5. CIR-IR spectra of ions in bulk solution versus ions at the solid/aqueous solution interface: (a) salicylate in D_2O at pD=4; (b) salicylate at the goethite/D_2O interface; (c) nitrate in H_2O; (d) nitrate at the goethite/H_2O interface at pH=3.5.

(bands at 1345 and 1356 cm⁻¹) in aqueous solution (**c**) with its corresponding vibrations when present at the goethite / aqueous solution interface (**d**). Since we know from adsorption studies that nitrate ions do not form chemical bonds with the surface of goethite, its absorption bands should be the same in both spectra, which is what we observe.

From studies of orthophosphate adsorption on the surface of goethite in aqueous suspensions [11] we noted that one type of adsorbate can form several type of complexes, depending upon suspension conditions. The spectra in Figure 6 are of phosphated goethite particles in aqueous suspensions for different values of pH and surface coverage. They show the existence of both a protonated and a deprotonated bridging bidentate complex. The relative integrated areas of the spectral bands for each of species type, normalized with respect to their relative extinction coefficients yields the value for the relative concentration of each species. The results of these studies, in combination with the interpretation of adsorption isotherms and electrophoretic mobility data, also allowed us to calculate (for the first time) the intrinsic pK values for the protonation of surface species, such as the bridging bidentate iron phosphate surface complex.

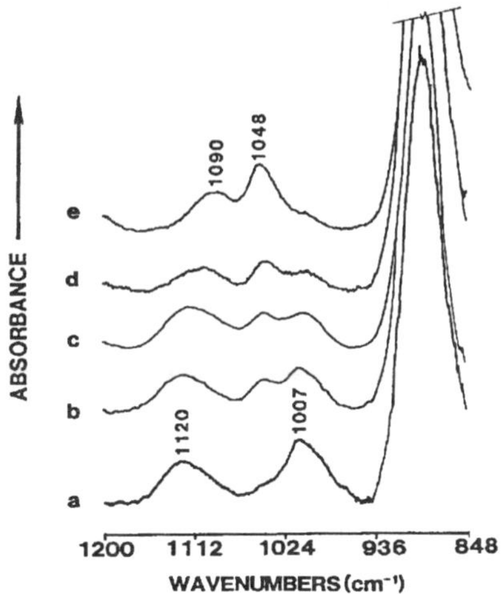

Figure 6. Influence of pH and surface coverage on the CIR-FTIR spectra of interfacial phosphate: (a) pH=4.0, 190 umol phosphate/g of goethite; (b) pH=5.0, 190 umol phosphate/g of goethite; (c) pH=6.0, 190 umol phosphate/g of goethite; (d) pH=5.0, 150 umol phosphate/g of geothite; (e) pH=8.4, 150 umol phosphate/g of goethite. (I = 0.01 $\underline{M}$ NaCl and solids concentration = 60 g/l.

We mentioned earlier that in ATR spectra of suspensions, the intensity of the absorption bands for particles may be dependent upon suspension conditions which lead to changes in the aggregation state of the particles. Although this complicates quantitative interpretations of data, it is ironically useful as a sort of nonspecific probe for studying aggregation. In a study on the interfacial chemistry of silicon nitride (Si_3N_4) particles in aqueous media, we used the ATR spectra as a nonspecific method to compare the state of aggregation of the particles under several suspension conditions, which would influence both the magnitude and the sign of the charge on the particles [12]. This phenomenon is illustrated in Figure 7. The spectra shown in **b** and **d** originate from clean silicon nitride, while those labeled **a** and **c** are from silicon nitride that had become contaminated with small particles of silica, which changed the mobility and aggregation behavior. The intensity of the adsorption bands in ATR spectra should also be applicable as an non-specific probe to study aggregation rates, for example.

Quantitative Characterization of Aqueous Suspensions

We then began the first steps in the making the transition of developing the ATR-FTIR spectroscopic technique from a qualitative into a quantitative technique for the study of

   Characterization Techniques for the Solid-Solution Interface

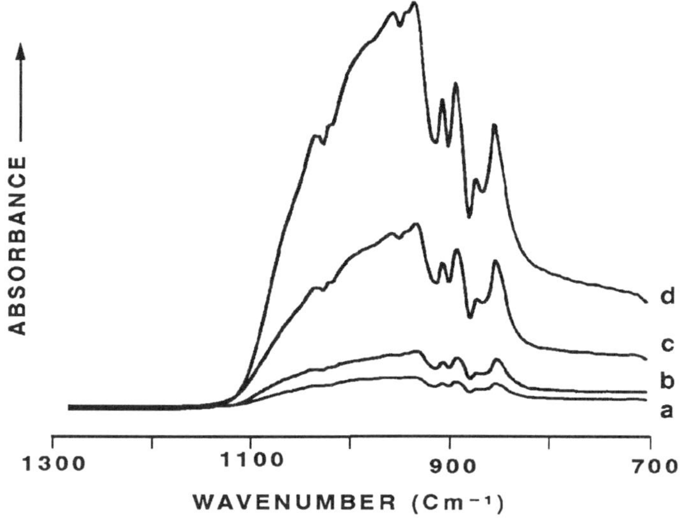

Figure 7. CIR-IR spectra of Si_3N_4 particles in H_2O for different pH values and pretreatment conditions. (a) and (c), Si_3N_4 contaminated with SiO_2 in H_2O at pH 7.0 and 4.0, respectively. (b) and (d) Si_3N_4 prewashed with base to remove SiO_2 at pH 9.3 and 4.0, respectively.

suspended phases [13]. We showed that, under a variety of suspension conditions, we could determine the spatial distribution of goethite particles in the region surrounding the IRE, and then compute the concentration of suspended particles in this region. We also were able to estimate values for the refractive indices of various goethite suspensions (in D_2O) at selected wavelengths during the course of computing the goethite concentrations.

Under suspension conditions where the ionic strength was low and the surface charge on the suspended particles was high, we calculated the refractive index (n) to be in the 1.4-1.5 range, which is significantly higher than the average value of 1.33 that one would predict from a simple weighted average of the contributions from pure deuterium oxide (98% by volume in the suspension) and pure goethite (2% by volume). Because we were able to determine the value for **n** at the frequency corresponding to the absorbance maxima for the goethite bands used in the concentration determination, and because there were no solvent bands in the region of the goethite bands, this difference in refractive index did not obstruct the determination of the values of goethite concentrations themselves. However, the disparity in **n** between reference and sample did contribute to strange band profiles in difference spectra in the spectral regions where strong absorbance bands due to the stretching and bending transitions for D_2O occur. An example of these shapes may be seen in Figure 2, in the region near 2000 cm⁻¹.

This is because, as we mentioned earlier, the shapes of absorption band profiles obtained

Characterization Techniques for the Solid-Solution Interface 161

in ATR spectroscopy are a function of many variables, including the refractive indices of both the sample being studied and the internal reflection element (IRE) that is used to probe the sample. Thus, absorption band intensities and band shapes contain contributions due not only to the extinction coefficient for the sample at various wavelengths, but also to the difference in refractive index between the sample and the IRE. In order to isolate the true absorbance spectrum for colloidal particles and the associated interfacial material, it is therefore necessary to consider differences in refractive index between the sample (suspension) and reference (supernatant). In order to achieve this, we would require a method to accurately determine the refractive indices for both the suspension and the supernatant samples over the entire spectral range, and be able to correct the data for these refractive index effects.

In fact, it is implicitly necessary to determine the dielectric properties of any sample for which quantitative interpretations of ATR spectra are to be made. Since ATR spectroscopy is based upon the phenomenon of total internal reflection, the absorption intensities and absorption band profiles obtained in an ATR experiment may be influenced by many variables that need not usually be considered in transmission experiments. The thickness of the sample, the refractive indices of the sample and the internal reflection element (IRE) which is used to study the sample, and the frequency and the angle of incidence of the probe radiation can all influence the spectrum obtained. Consequently, it is clear that one must know or be able to determine the values of both the extinction coefficient (**k**), which is related to the absorption coefficient, α, and the refractive index (**n**) of any sample, as a function of wavelength, in order to accurately study vibrational modes or to perform quantitative analyses. In order to remove spectral artifacts in spectra due to refractive index differences between reference and sample, it is necessary to account for these differences by adjusting the raw spectra prior to subtraction. Thus, it is first necessary to determine the refractive indices of both the reference and the sample. Fortunately, the average refractive index of any material which is uniform in concentration, behaves as a thick film, and is transparent in the infrared region of the spectrum, can be determined by analyzing spectra collected at different angles of incidence.

During the past decade, ATR-FTIR spectroscopy has been used increasingly for the determination of optical constants (extinction coefficients and indices of refraction) in the infrared region of the spectrum [14-16]. We have adapted this methodology for determining the variation of the refractive index and extinction coefficients for suspensions with the frequency of the IR radiation.
The extinction coefficient is related to the absorption coefficient, α (introduced in equations 1 and 2), by the expression:

$$k = \frac{\alpha c}{4\pi v}$$

(4)

where v is the wavenumber. The extinction coefficient is thus dimensionless and describes the amplitude of the absorption for pure materials. Optical calculations require

     Characterization Techniques for the Solid-Solution Interface

the use of the dimensionless k rather than α, in constructing an important relationship with the refractive index, n, which will ultimately enable us to perform quantitative ATR-FTIR spectroscopy. It is thus useful to rewrite equation (2) in terms of k rather than α:

$$a = \frac{2Nkn_{21}\cos\Theta[(3 + n_{21}^2)\sin^2\Theta - 2n_{21}^2]}{(1 - n_{21}^2)[(1 + n_{21}^2)\sin^2\Theta - n_{21}^2](\sin^2\Theta - n_{21}^2)} \tag{5}$$

It is important to note the intimate relationship between the refractive index and the extinction coefficient as expressed by the Kramers-Kronig transform [17]:

$$n(\nu_c) - n_\infty = \int_0^\infty \frac{k(\nu)}{\nu^2 - \nu_c^2} d\nu \tag{6}$$

where n is the baseline refractive index (anchor point). (The Kramers-Kronig transform can actually be rewritten as a Fourier-type transform and readily solved by computer, if the value for n is known or can be determined). Recently, we have developed a methodology for determining the average refractive index of suspensions using variable angle ATR-FTIR spectroscopy [18]. The method uses the absorbance values for weakly absorbing bands for water in the suspension and the supernatant which have been collected at different angles of incidence. The values we compute are thus averages of contributions from both "bulk" and "interfacial" molecules. We can then use these values along with the Kramers-Kronig transform to separate the contribution of the refractive index from absorbance bands in spectra for suspensions. Finally, we can use a rearranged form of equation (5) to obtain the true spectrum for suspended particles and "interfacial" water and allow us to determine the extinction coefficients at various wavelengths.

Given the complexities involved in quantitatively interpreting spectra, we have found that creating suspension conditions which follow the thick film model can greatly simplify our efforts. Generally, concentrated (>2% solids by volume), well dispersed suspensions with ionic strength values less than 0.01 molar fall into this category. Systems which follow the "thick" film model offer several advantages over the other models. First of all, we can assume that such systems are homogeneous at a scale of homogeneity of the order of the probe radiation (i.e. microns, in this case). Thus, the concentrations of the various components present in the sample, such as suspension particles, "interfacial" molecules, and "bulk" water molecules, can be assumed to be independent of the distance from the IRE. The concentrations of all types of molecules in the suspension sample are, therefore, independent of the angle of incidence with which the infrared beam strikes the internal surface of the IRE. Furthermore, the actual value for the film thickness need not be known. In addition, the average refractive index of the system can also be assumed to be independent of the angle of incidence.

Systems which follow the thin film model will also readily lend themselves to quantitative analysis. However, we have found that the model is generally useful for describing only very dilute suspensions in which the IRE and the particles are of opposite charge. We have discussed this situation elsewhere [13], but due to its limited range of applicability, we will not discuss it further here.

As we mentioned earlier, we have noticed significant increases in the refractive indices for suspensions containing highly charged, well dispersed particles. Because we believe it to be unlikely that simple contributions from the solid particles themselves are the sole source of these increases, we believe them to be due, in part, to structuring of the water by the charged surface. We will discuss this further, along with implications for ceramic membranes, in the next section. We will also demonstrate that a good method for studying the interfacial properties of metal oxides under all solution conditions involves sintering the individual oxide particles to form ceramic membranes that meet thick film criteria. Since these sintered films are rigid, the arrangement of the oxide particles in the film is not subject to changes due to solution parameters such as pH and ionic strength.

APPLICATION TO CERAMIC MEMBRANES

Porous ceramic thin films constitute an intriguing new class of materials with potential for a wide range of industrial applications--as separation devices (in ultrafiltration, reverse osmosis systems, and membrane reactor/separators), as catalysts for the photodegradation of organic contaminants in air and in aqueous solution, as elements for the photovoltaic conversion of energy, and as components of extremely thin batteries. One procedure for making the particles that are suitable for producing such films is known as the sol-gel process. In the sol-gel process [19], covalent metal alkoxides are hydrolyzed to produce small metal oxide particles. Polycrystalline porous metal oxide films may be produced by applying wet films of these suspended oxide particles to a supporting medium and sintering or calcining the films at appropriate temperatures. The specific crystalline forms which are obtained are determined largely by the calcination conditions. Oxide films with thicknesses in the range of 0.01-10u can be prepared from a wide variety of precursors using this technology. Selection of the starting materials, coupled with a knowledge of how to manipulate fabrication parameters, allows one to control not only the chemical composition of the film, but also its optical, electronic, catalytic and surface properties. Research in our laboratory and elsewhere has demonstrated that one can fashion such films having characteristic dimensions of less than 50Å and very large accessible internal surface areas (up to several hundred m^2/g) that meet thick film criteria. The pore sizes in these films can be tailored to control diffusion rates or to effect size-specific reaction chemistry. To date, we have produced films of silica, titania, alumina, zirconia, and various iron oxides, representing a wide range in surface properties and hydrophilicity.

We are presently developing two methods for studying ceramic films using ATR-FTIR spectroscopy. In the first approach, we attach the films directly to the IRE. In this case, the film actually becomes part of the IRE, and can be treated as part of the background, which is very advantageous. The main disadvantage of this method is that, in addition to the IRE considerations mentioned above for studying suspensions, we must now also consider sample adhesion properties, compatibility between the sample and IRE, and thermal stability of the IRE (if the sample is to be fired on the IRE). Following immersion of the supported dry film in water, we can then we can directly generate spectra for the pore contents (both adsorbed and non-adsorbed molecules) and automatically eliminate contributions from bands due to the metal oxide crystallites themselves. The water surrounding the film can be adjusted to different pH and ionic strength values to change the nature of the electrical double layer in the pores. We can then record spectra, and determine the refractive index and extinction coefficient of the pore contents. From this data, we should be able to relate the surface potential, thickness of the double layer and size of the pore to the structure of water in these confined spaces.

In the second approach, we attach the ceramic film to an optically flat substrate (such as silica) and press it against a stationary IRE. The main advantage to this approach is that the substrate is not required to have special optical properties (as with IREs), and is therefore much less expensive. The main disadvantage is that a thin film of nonadsorbed water ("bulk" water) will often be present between the IRE and the sample, which can complicate interpretations of spectra.

In the development of ceramic membrane technology, an understanding of water structure near an interface is acutely important. Over the past five years, we have been developing a broad array of ceramic membrane filters for a variety of applications. Recently, we have produced membranes having pore diameters below 1.5 nm, for reverse osmosis applications [20,21]. Permeability tests suggest that the structure of interfacial water may affect the flow of aqueous solutions through these small pores.

We noted in previous work [7] that the hydrogen bond distribution in water present in the metal oxide/liquid interface is considerably different from that for bulk water. As mentioned earlier, our ATR-FTIR spectroscopy studies [13] performed on aqueous suspensions of goethite (α-FeOOH) reveal that, under certain solution conditions which led to highly charged particles, the refractive index of the suspension is 10-15% higher than that predicted. Other researchers have suggested this as a result of modeling studies [22].

Indeed infrared spectroscopy will provide information as to whether the surface promotes or diminishes the degree of hydrogen bonding of the vicinal water. The profile of absorption bands in infrared spectra of such water reflects the distribution of bond lengths and angles of the water molecules seen by the radiation. Although it is not possible at the present time to calculate the actual values for bond lengths and angles of these molecules from the absorption frequencies, one can formulate a picture of the surface configurations of these molecules by comparing their absorption frequencies with those of water molecules in a single crystal for which the coordinates of the atoms are known.

Both the extinction coefficient and the refractive index can provide information that is useful in the comparison of "interfacial" water with "bulk" water. The extinction coefficient can provide information about oscillator (bond) strength; shifts in distributions of the extinction coefficient as a function of frequency for water in different types of samples can yield information on bond energies and the extent of hydrogen bonding. From the refractive index values we can also determine the dielectric constant and the polarization of water molecules in these pores.

The refractive index can be used as a yardstick for comparing the relative densities or "packing" of water molecules in different sample types, since the refractive index, on a molecular level, is related to the number of atoms per unit volume. Thus, significant changes in refractive index for different solution or suspension conditions may imply changes in density or structure involving at least a fraction of the water molecules in the system. These data, along with data from other types of studies, can allow one to formulate models to describe the configuration of water molecules at the interface and explain the roles of surface charge, chemisorption, and hydrogen bonding on the structure of "interfacial" water. Because we can vary the pore sizes in ceramic membranes by changing fabrication variables, we should be able to find a pore diameter below which most or all of the water is "interfacial."

CONCLUSIONS

The investigation of aqueous suspensions and ceramic membranes using ATR-FTIR spectroscopy presents many challenges to the analytical chemist. Several factors, including interactions between particles, sample-IRE interactions, the strong absorbance of IR radiation by water, and overall optical considerations, must be accounted for. However, by manipulating the sample conditions, and even changing the sample form through sintering, we are able to minimize the problems caused by these factors and even turn some of the problems into advantages. The results of this research could potentially have significant applications in the study of filtration mechanisms, biological membrane processes, studies of gelation, the behavior of latex systems, and the transport of water through paint films, and other processes where understanding of the nature of the electric double-layer is essential.

REFERENCES

1.	Seki, H., K. Kunmatsu and W. G. Golden, <u>Appl. Spectrosc.</u>, <u>39</u>, 437. 1985.

2.	Tejedor-Tejedor, M. I. and M. A. Anderson, <u>Langmuir</u>, <u>2</u>, 203. 1986.

3.	Zeltner, W.A., E. Yost, M. Machesky, M.I.Tejedor-Tejedor and M.A. Anderson. in <u>Geochemical Processes at Mineral Surfaces</u>, ACS Symp. Ser. <u>323</u>, 142. 1986.

4.	Harrick, N. J. and F. K. du Pre', <u>Appl. Optics.</u>, <u>5</u>, 1739. 1966.

5.	Harrick, N. J., and F. M. Mirabella, Jr., <u>Internal Reflection Spectroscopy: Review and Supplement</u>, Harrick Publishing Company, Rahway, NJ. 1983.

6.   Harrick, N., <u>Internal Reflection Spectroscopy</u>, Wiley and Sons, New York. 1967.

7.   Tejedor-Tejedor, M.I., and M.A. Anderson, <u>Langmuir</u>, 2, 203. 1986.

8.   Tejedor-Tejedor, M.I., E.C. Yost, and M.A. Anderson, <u>Langmuir</u>, 6, 979. 1990.

9.   Tejedor-Tejedor, M.I., E.C. Yost, and M.A. Anderson, <u>Langmuir</u>, 8, 525. 1992.

10.   Tejedor-Tejedor, M.I. and M.A. Anderson, <u>Langmuir</u>, 602. 1990.

11.   Tunesi, S. and M. A. Anderson, <u>Langmuir</u>, 8, 487. 1992.

12.   Zeltner, W.A., M.I. Tejedor-Tejedor, and M.A. Anderson, "CIR-FTIR Studies of the Solid/Liquid Interface of UBE Silicon Nitride Powders During the Aging Process," to be published.

13.   Tickanen,L.D., M.I. Tejedor-Tejedor, and M.A. Anderson, <u>Langmuir</u>, 7, 451. 1991.

14.   Bertie, J. E., M. K. Ahmed, and H. H. Eysel, <u>J. Phys. Chem.</u>, 93, 2210. 1989.

15.   Braue, E. H. and M. G. Panella, <u>Appl. Spectrosc.</u>, 41, 105. 1987.

16.   Goplen, T. G., D. G. Cameron, and R. Jones, <u>Appl. Spectrosc.</u>, 34, 652, (1980).

17.   Lipson, S. and H. Lipson, <u>Optical Physics</u>, Cambridge Press, New York. (1981).

18.   Tickanen, L.D., M.I. Tejedor-Tejedor, and M.A. Anderson, "The Concurrent Determination of Optical Constants and the Kramers-Kronig Integration Constant Using ATR-FTIR Spectroscopy, submitted to <u>Applied Spectroscopy</u>, Feb. 1992.

19.   Yoldas, B. E., <u>Amer. Ceram. Soc. Bull.</u>, 54, 3. 1975.

20.   Gieselmann, M. J., M. A. Anderson, M. D. Moosemiller, and C. G. Hill, Jr., <u>Separat. Sci. Technol.</u>, 23 [12 & 13], 1695. 1988.

21.   Anderson, M.A., M.J. Gieselmann, and Q. Xu, <u>J. Membr. Sci.</u>, 39, 243. 1988.

22.   James, R.O., and T.W. Healy, <u>J. Colloid Interface Sci.</u>, 40, 42, 53, 65. 1972.

CHARACTERIZATION OF THE METAL OXIDE-POLYACRYLIC ACID COMPOUNDS BY FOURIER TRANSFORM INFRARED SPECTROSCOPY

José M. Saniger,* H. Hu*, Víctor M. Castaño,† Juventino García-Alejandre‡
*Centro de Instrumentos, Universidad Nacional Autónoma de México (UNAM), Apdo. Postal 70-186, México, D.F. 04510
†Instituto de Física, UNAM, Apdo. Postal 20-364
‡Facultad de Química, División de Estudios de Posgrado, UNAM

ABSTRACT

The criteria, still under revision, for the characterization by Fourier Transform Infrared Spectroscopy (FT-IR) of the molecular arrangement of the metal-carboxylate complex, are applied to divalent and trivalent metal oxide-polyacrylic acid compounds. The results show that the so-called bidentate chelating or bridging structure are the most likely to occur in this case. The applicability of the technique to other systems is also discussed.

INTRODUCTION

Polyacrylic acid - metal oxide (PAA-MO) compounds, which form a kind of ionic polymer cement, have been studied for several years, due to their surgical and dental uses as biocompatible ceramics (1-5), and having potential applications as composites materials. PAA-MO compounds are easily prepared at room temperature; they have a very fast rate of hardening, good mechanical properties (6-7) and good adhesion to different engineering materials (8). Besides their practical applications, they are of special interest for basic research due to the presence of organic and inorganic phases chemically bonded, which results in a cross-linked glassy material, whose properties can be modified by changing adequately the nature of the metal oxides and the polymer. The infrared spectroscopy has been used to show the existence of a chemical interaction between the PAA and metal oxides and even to propose possible molecular arrangements in the polyelectrolyte cement (9-11). In this work we present a study by FT-IR of different compounds of PAA with divalent and trivalent metal oxides and a discussion about their possible molecular conformations.

MATERIALS AND METHODS

The samples were prepared by adding, in the stoichiometric proportion, the powders of the different metal oxides to an aqueous solution of PAA (25% wt., 90,000 average molecular weight) and manually stirring the mixture for two or three minutes. After that, a viscous substance with a high water content was obtained, and was cured for 72 hours under room conditions. The metal compounds used were MgO, CaO, ZnO, CuO, Nd_2O_3, CoO(OH) and a polynuclear species of aluminum oxohydroxide. Powders of reactive grade MgO, CaO, Nd_2O_3 and ZnO were used as-received. CuO was prepared by thermal decomposition of malachite, $Cu_2CO_3(OH)$, and the black powder obtained was identified as CuO by X-ray diffraction and it had higher porosity and chemical reactivity as compared to the commercial oxide which hardly reacts at our experimental conditions. CoO(OH) was prepared by precipitation of $Co(OH)_2$ from a mixture of aqueous solution of $COCl_2-2H_2O$ and NaOH. The precipitate was air heated at 150°C in order to oxidize it until CoO(OH) was formed (12). The aluminum oxohydroxide was precipitated by adding a $O.1N$ NH_4OH solution to a $O.1N$ $Al_2(SO_4)_3 \cdot 18H_2O$ solution, until a pH of 7 was obtained. At this point the precipitate was filtered and dried, before using it for the reaction with the PAA. It is normally accepted (13-15) that this precipitate is formed by polynuclear ionic species with a composition of $(Al_{13}O_4(OH)_{24}(H_2O)_{12})^{+7}$ which is electrically compensated by the co-precipitation of anionic species, basically SO_4^{2-}. The precipitate reacts easily with the PAA solution to form the corresponding aluminum salt.

The characterization of the PAA-MO compounds was performed by comparing the FT-IR spectra of the unreacted PAA with the reacted mixtures of PAA and MO, prepared as was explained above. Unreacted PAA films were prepared by dissolving a drop of the PAA solution in methanol, and by depositing the mixture onto a KBr window with further evaporation of the solvent. The cured PAA-MO compounds were powdered in an agate mortar and mixed with powdered KBr in a proportion of 1/100. The resulting mixture was compacted until a sample disc with the adequate transparency was obtained. The spectra were registered between 4000 and 600 cm^{-1} in a Nicolet 5PC FT-IR spectrometer.

RESULTS AND DISCUSSION

The spectra of the unreacted PAA and its sodium salt are shown in Figure 1, whereas Table 1 presents the frequencies and the assignments of the main absorption bands of both spectra.

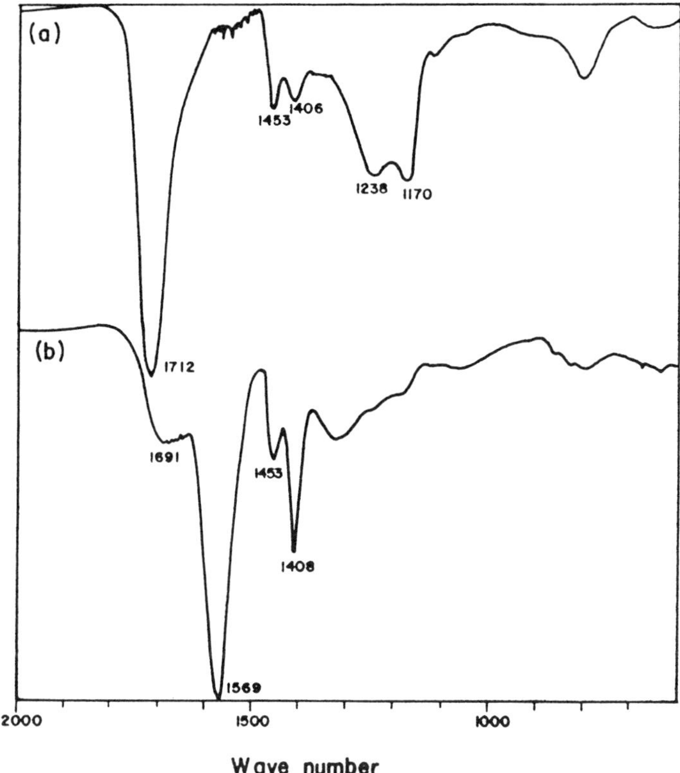

Figure 1. FT-IR spectra of PAA (a) and its sodium salt (b).

The strong absorption band of the PAA at 1712 cm^{-1} is very close to the stretching vibration of the carbonyl groups, $>C=O$, of two carboxylic groups dimerized via hydrogen bond (16). The weaker bands at 1453 cm^{-1} and 1406 cm^{-1} can be associated with the bending vibrations of $-CH_2-$ and $-CH_2-CO-$ groups, respectively. Finally, the bands at 1235 and 1170 cm^{-1} may be related to the coupling between in-plane O-H bending and C=O stretching of the carboxylic dimer. The existence of free polyacrylic acid with a high grade of dimerization between neighboring carboxylic groups is deduced from this spectrum.

Compound	Frequency (cm^{-1})	Assignment
PAA	1712	υ (C=O)
	1453	δ (CH$_2$)
	1406	δ (CH-CO-)
	1238	see text
	1170	see text
PAA-Na	1569	υ_{as} (- C $\overset{O}{\underset{O}{}}$)
	1453	δ (CH$_2$)
	1408	υ_{as} (- C $\overset{O}{\underset{O}{}}$); δ (CH-CO-)

Table 1. The frequencies and the assignments of the main absorption bands of FT-IR spectra of the unreacted PAA and its sodium salt.

When the PAA is neutralized with a NaOH solution, the sodic salt is formed, in which the carboxylate group, [COO]$^-$, is ionically bonded to the Na$^+$ cation (17). Consequently, the 1712 cm^{-1} band of the carbonyl group is split in two new bands (1569 cm^{-1}, 1408 cm^{-1}) due to the asymmetric and symmetric stretching vibrations of the carboxylate anion formed. The band at 1453 cm^{-1}, associated with the bending vibrations of -CH$_2$- group remains unchanged since the -CH$_2$- group is not involved in the reaction. Nevertheless, there is an enhancement in the intensity of the band at 1406 cm^{-1}, although the -CH$_2$-CO- group are not involved in the reaction. We assume, according to Bardet et al. (10), that this phenomenon is due to the superposition of two bands; one the bending vibration of the -CH$_2$-CO- group, and the other, the symmetric stretching vibrations of the carboxylate anion. From here, it follows that the bands of the [COO]$^-$, asymmetric and symmetric stretching, have frequencies of 1569 and 1408 cm^{-1} respectively,

   Characterization Techniques for the Solid-Solution Interface

with a difference of 161 cm^{-1}. This will be a very important reference in the following discussion.

The spectra of all the others PAA-MO compounds are very similar to the spectrum of the sodic salt of the PAA. So, it is possible to conclude that an analogous reaction has occurred in all these cases.

Infrared spectroscopy has been widely used to propose molecular arrangements of the carboxylate compounds, by using empirical correlations between the carbon-oxygen stretching frequencies of the carboxylate ions and the nature of their coordination (18). However, there is no general agreement for the criteria that must be followed to correlate the values of the stretching frequencies of carboxylate and the kind of ligand formed. In this work, we apply the criteria used by Deacon and Phillips (18) and Nakamoto (17), for the study of the structures of the aceto-metal complex. They basically consider the difference between the asymmetric and symmetric stretching frequencies, $\Delta\upsilon$, of the carboxylate group, to discern among different possible structures.

A carboxylate ion [R-COO]$^{-}$, can coordinate to metals fundamentally in three ways, as monodentate ligand (I), bidentate chelating ligand (II), and bidentate bridging (III), as shown in Figure 2.

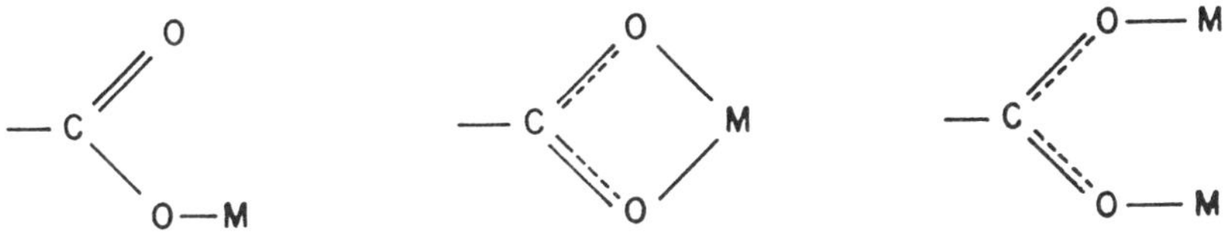

O   O   O — M

— C   — C   M   — C

O — M   O   O — M

(I)   (II)   (III)

monodentate   bidentate chelating   bidentate bridging

Figure 2. Three different types of coordination of carboxylate ion to metals

In addition to coordination derivatives, ionic metal carboxylates are also possible, but in these compounds the cations are constrained to remain in the vicinity of the carboxylates anion by a general electrostatic field and there are no specific binding sites. Carboxylic alkali metal salts are examples of ionic carboxylates.

As we discussed before, the $\Delta\upsilon$ for PAA sodic salt is 161 cm^{-1}. With this value as reference, when a carboxylate complex is formed having a $\Delta\upsilon$ higher than 200 cm^{-1}, then the more probable structure is the monodentate. On the other hand, when $\Delta\upsilon$ is lower than 105 cm^{-1}, the bidentate chelating ligand is the more likely arrangement. For the values of $\Delta\upsilon$ between 105 and 200 cm^{-1} there is no general agreement, and the less risky assumption is that either bidentate bridging or chelating structure could be formed (17-18). Table 2 presents the values of the υ_{as}, υ_s and $\Delta\upsilon$ for the different PAA-MO under study, together with the proposed molecular arrangement for each one.

Compound	υ_{as} (cm^{-1})	υ_s (cm^{-1})	$\Delta\upsilon$ (cm^{-1})	Proposed Structure
PAA-Na	1569	1408	161	ionic
PAA-Mg	1558	1412	146	ionic
PAA-Ca	1556	1415	141	ionic
PAA-Zn	1570	1410	160	(II), (III)
PAA-Cu	1562	1408	154	(II), (III)
PAA-Co	1552	1409	143	(II)*, (III)
PAA-Nd	1544	1414	130	(II)*, (III)
PAA-Al	1620	1460	160	(II), (III)*

*considered to be the more likely structure.

Table 2. υ_{as}, υ_s and $\Delta\upsilon$ values for different PAA-MO complexes.

The assumption of the ionic character of Mg and Ca salts of PAA is based on the chemical characteristic of these cations, principally the high ionic percentage of their salts. Besides, the mechanical properties of the two compounds in water are quite different from the other listed in Table 2; they are much more elastic and they absorb and desorb in a reversible way a large amount of water (7). These

   Characterization Techniques for the Solid-Solution Interface

properties are consistent with the nonexistence of a localized site for the bonding, which is a characteristic of the ionic bond.

The remainder cations, normally form carboxylate coordination compounds (17-18) and therefore its bonding mode can be discussed with the exposed criteria. Divalent zinc and copper compounds have very close values for ν_{as}, ν_s and $\Delta\nu$ with those of the sodium salt, and then both the bidentate bridging and chelating structures must be equally considered.

Something similar occurs in trivalent cobalt and neodimium compounds, but with the difference that ν_{as} is significantly lower than that of the sodic salt. In this case some authors (19-20) consider that the bidentate chelating structure has higher probability. The aluminum compound also has a $\Delta\nu$ very close to the sodium salt, but, conversely, now the ν_{as} is clearly higher than that of the ionic sodium compound, and then the bidentate bridging structure can be considered preferentially.

CONCLUSIONS

From the results obtained in the current work, it can be concluded that the PAA-MO polyelectrolytic cements do not consist of a physical mixture of the components, but it is clear that a chemical bond forms between the polymer and the various metals.

Alkaline earth metals (Mg and Ca) form, as could be expected, ionic compounds with mechanical properties quite different from the other divalent transition metals tested (Zn and Cu). This opens the possibility of designing the mechanical properties of these materials by changing adequately the ionic/covalent ratio of their polycarboxylate-metal chemical bond.

In our case, it is not possible, at the moment, to characterize precisely the structure of all the carboxylate-metal ligand under study, by using only the FT-IR spectroscopy, because the empirical correlations between $\Delta\nu$ and ligand structure are still under discussion, especially when the values of $\Delta\nu$ for the compound and the ionic salt are close. In our opinion, the study of other possible empirical correlations of the structure with mechanical or thermal properties of the compounds will be useful to advance the characterization of these materials. Knowing that X-ray diffraction is not applicable here, because of the amorphous conditions of these compounds, nuclear magnetic resonance (NMR) appears as the other analytical technique clearly useful for our purpose.

ACKNOWLEDGEMENT

The presentation of this work has been partially supported by CONACyT (Consejo Nacional de Ciencia y Tecnología).

REFERENCES

1. D.C. Smith, *Brit. Dent. J.*, **125**, 381 (1968).
2. A.D. Wilson and B.E. Kent, *J. Appl. Chem. Biochtechnol.*, **23**, 811 (1971).
3. J.A. Barton, G.M. Brauer, J.M. Antonucci, and M.J. Rauey, *J. Dent. Res.*, **54**, 310 (1975).
4. K.A. Wood and A.G. Reader, *Br. Polym. J.*, **8**, 131 (1976).
5. L.L. Hench and D.R. Ulrich, *Ceramics Chemical Processing*, John Wiley and Sons, New York, 1986.
6. A. Padilla, A. Vázquez, and V.M. Castaño, *J. Mater. Res.*, **6**, 11, 2452 (1991).
7. H. Hu, Ph.D. Thesis, Universidad Nacional Autónoma de México, UNAM, 1992.
8. T. Sugama, L.E. Kukacka, and N. Carciello, *J. Mater. Sci.*, **19**, 4045 (1984).
9. S. Crisp, H.J. Prosser, and A.D. Wilson, *J. Mater. Sci.*, **11**, 36 (1976).
10. L. Bardet, G. Gassanas-Fabre, and M. Alain, *J. Molecular Structure*, **24**, 153 (1975).
11. H. Hu, J.M. Saniger, J. García-Alejandre, and V.M. Castaño, *Mater. Lett.*, **12**, 281 (1991).
12. N.N. Greenwood and A. Earnshaw, *Chemistry of the Elements*, Pergamon Press, 1984.
13. C.F. Baes and R.E. Mesmer, *The Hydrolysis of Cations*, Wiley, New York, 1976.
14. S.L. Nail, J.L. White, and S.L. Hem, *The Journal of Pharmaceutical Sciences*, **65**, 1188 (1976).
15. R.J. Moolenaar, J.C. Evans, and L.D. McKeever, *J. Phys. Chem.*, **74**, 3629 (1970).
16. K. Nakanishi, *Infrared Absorption Spectroscopy*, Holden Day Inc./Nankodo Co. Ltd., Tokyo, 1969, p. 39.
17. K. Nakamoto, *Infrared and Raman Spectra of Inorganic and Coordination Compounds*, Wiley, New York, 1978, pp. 232-233.
18. G.B. Deacon and R.J. Phillips, *Coordination Chemistry Reviews*, **33**, 227 (1980).
19. A.V.R. Warrier and R.S. Krishnan, *Spectrochim. Acta*, **27**, 1243 (1971).
20. K.S. Patel, J.A. Faniran, and A.Earnshaw, *J. Inorg. Nuc. Chem.*, **38**, 352 (1976).

ANALYSIS OF BIOACTIVE GLASS INTERFACIAL REACTIONS USING FOURIER TRANSFORM INFRARED REFLECTION SPECTROSCOPY

Guy LaTorre and Larry L. Hench, University of Florida, Advanced Materials Research Center, One Progress Blvd., #14, Alachua, FL 32615

INTRODUCTION TO FTIR TECHNIQUES AND THEORY

Background From Our Laboratory

Infrared reflection spectroscopy has been used to study glass-environment interactions for twenty years. Sanders, et al. used a simple grating spectrometer and home made specular reflection stage to measure the kinetics of alkali loss from binary alkali-silicate glasses in contact with water.[1-6] Silica repolymerization of the glass surface was also observed with the IR reflection (IRRS) measurements. Many variables that affect glass-solution reactions were studied using the IRRS method including: effects of surface roughness,[4] ratio of network modifier to network former,[5] ratio of glass surface area to volume of reacting solution (SA/V),[3,8-13] atmosphere effects,[3,13] composition of reactant solution,[10,11] mixed alkali effect,[10] alkali-alkaline earth effect,[11] and effects of surface stabilizing[12,13] polyvalent cations.[14-16] Refer to references 14-16 for discussion of studies from other laboratories.

The IRRS method has been used extensively to understand the composition dependence of nuclear waste glasses[17] and nuclear waste glass-repository environment interactions.[18] IRRS analyses have also been invaluable in understanding the behavior of bioactive glasses and glass-ceramics.[19-25] Bioactive glasses are now in clinical use and FTIR methods are the basis for quality assurance (QA) and quality control (QC) tests to ensure their reliability for use in humans.[26] A new generation of bioactive glasses made by sol-gel processing has been developed using FTIR characterization methods.[27] Therefore, this chapter uses these important materials as the basis for discussing the FTIR method of analyzing glass-solution interactions.

The Theory of IR Adsorption

The infrared region of the spectrum ranges in wavelength from 12,800 inverse centimeters (cm^{-1}) to 10 cm^{-1}, i.e., 0.78 to 1000 μm. This range of frequencies is generally subdivided into three regions: the near IR from 12,800 to 4,000 cm^{-1}, the mid IR from 4,000 to 200 cm^{-1} and the far IR from 200 to 10 cm^{-1}. The region most often used in analytical IR analysis for most organic and inorganic compounds is the mid infrared. In order for IR absorption to occur in this region a molecule must exhibit a net change in dipole moment as a result of its vibrational or rotational motion. The

Characterization Techniques for the Solid-Solution Interface

dipole moment is dependent on the magnitude of the charge difference and the interatomic distance between two atoms that form a bond. When the frequency of incident radiation matches the frequency of vibration of a bond in a molecule the absorption of radiation occurs. Vibrational transitions occur in frequency ranges from 13,000 to 400 cm^{-1} and are the most useful for characterizing organic and inorganic compounds. Rotational transitions occur in frequency ranges of <100 cm^{-1} and are highly restricted in liquid and solids. Therefore, they are of little interest to spectroscopists trying to analyze most materials.

Vibrational transitions can fall into two categories; stretching vibrations which occur when there is a change in interatomic distance between two atoms, and bending vibrations which occur when there is a change in bonding angle between two bonds sharing an atom. It is both types of vibrational transitions that the IR spectroscopist uses to characterize solids and liquids.

The basis for quantitative absorption or transmission spectroscopy is described by Beer's Law

$$A = \log 1/T = abc \qquad (1)$$

where: A = the absorbance
T = the transmittance
a = sample absorptivity
b = sample thickness or path length
c = concentration of absorptive component in the sample

This relationship is useful only when applied to transmission or absorption measurements where these variables can be controlled or determined. Deviations from this ideal can occur as a consequence of the manner in which the measurements are made, as is the case with reflection spectroscopy. When infrared radiation is reflected from a smooth flat surface, as in specular reflectance (Fig. 1A), some of the radiation penetrates into the surface and some is reflected from the surface. This depth of penetration, or the pathlength, is a function of the angle of the incident beam, the refractive index of the sample, the wavelength of radiation and the smoothness or flatness of the sample surface. A portion of the radiation penetrates into the surface and an absorption of radiation occurs before reflection occurs. The result of this interaction is a spectrum that contains both an absorption component and a refractive index component for the sample being analyzed. When the sample surface is irregular or rough there is an increase in the amount of scattered or diffused energy as a result of surface scattering and internal reflection or attenuation of the infrared radiation (Fig. 1B). Because of these kinds of interactions it is not always possible to account for all of the variables with respect to Beer's Law and quantitative measurements using reflection spectroscopy become difficult and in some cases impossible. There are ways of reducing these variables for solid samples, such as polishing the sample surface or controlling the angle of the incident beam, but in the case of irregular surfaces such as powders this cannot be done and qualitative information for the sample can only be derived. This qualitative information can also be very useful to the spectroscopist. However, analysis of diffuse reflection is possible with a proper stage and can yield quantitative data, depending upon experimental conditions. An example of quantifying

   Characterization Techniques for the Solid-Solution Interface

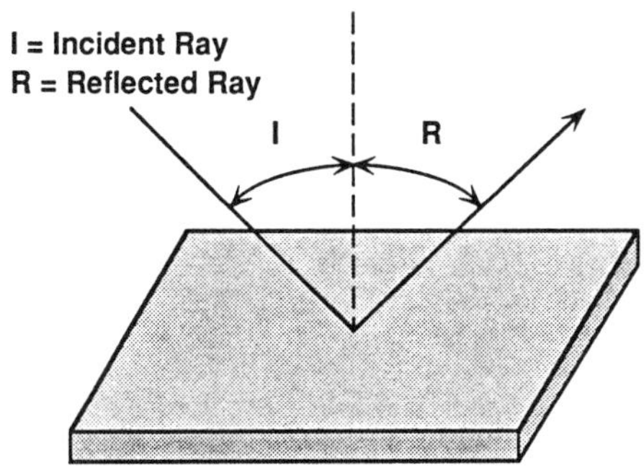

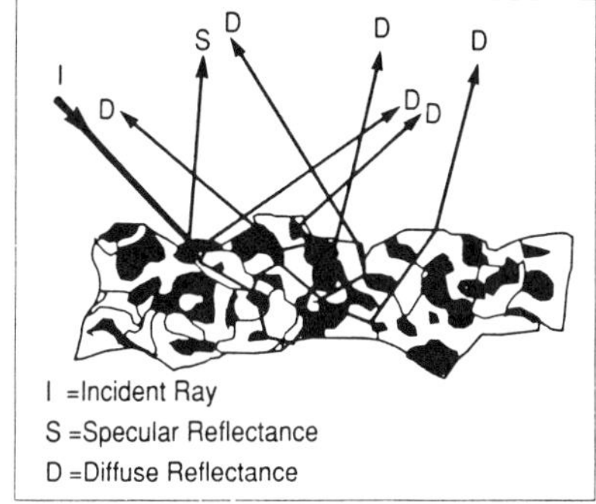

Figure 1. The different beam paths for specular (A) versus diffuse (B) reflectance.

surface reaction kinetics using diffuse reflection is discussed later. The reader should refer to refs. 28-32 for a more extensive discussion of these complex interactions.

BIOACTIVE GLASS AND SOLUTION REACTIONS

Specific compositions of SiO_2-CaO-Na_2O-P_2O_5 glasses form a bond with living tissues and are called bioactive glasses or, when crystallized, bioactive glass-ceramics.[24] Figure 2 summarizes the compositional dependence of bioactive glasses. The compositional dependence (in weight %) of bone-bonding and soft-tissue bonding for the Na_2O-CaO-P_2O_5-SiO_2 glasses is illustrated in Fig. 2. All glasses in Fig. 2 contain a constant 6 weight % of P_2O_5. Compositions in the middle of the diagram (Region A) form a bond with bone. Consequently, Region A is termed the bioactive-bone-bonding boundary. Silicate glasses within Region B (such as window, bottle glass or microscope slides) behave as nearly inert materials and elicit a fibrous capsule at the implant-tissue interface. Glasses within Region C are resorbable and disappear within 10 to 30 days of implantation. Glasses within Region D are not technically practical and so have not been tested as implants. The 45S5 Bioglass® composition discussed below lies in the middle of Region A.

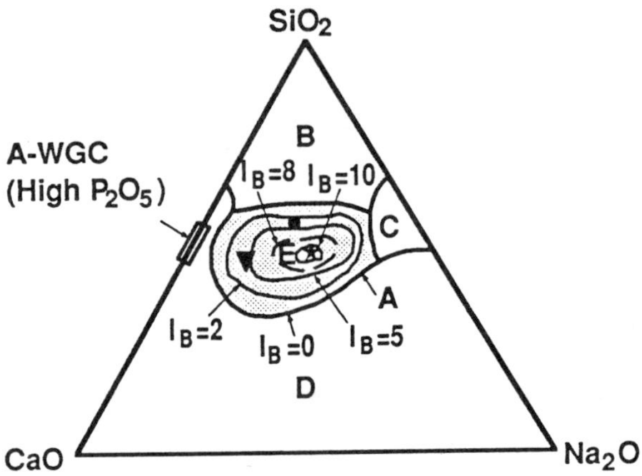

Figure 2. Compositional dependence (in weight percent) of bone bonding and soft-tissue bonding of bioactive glasses and glass-ceramics. All compositions in region A have a constant 6 wt% of P_2O_5 content (see Table 1 for details). Region E (soft-tissue bonding) is inside the dashed line where $I_B > 8$(★) 45S5 Bioglass®, (▼) Ceravital®, (●) 55S4.3 Bioglass®, and (---) soft-tissue bonding; $I_B = 100/t_{0.5bb}$.

Table 1. Reaction Stages of a Bioactive Implant

Stage	Reaction
1	Rapid exchange of Na^+ or K^+ with H^+ or H_3O^+ from solution: $Si\text{-}O\text{-}Na^+ + H^+ + OH^- \rightarrow Si\text{-}OH^+ + Na^+(solution) + OH^-$ This stage is usually controlled by diffusion and exhibits a $t^{-1/2}$ dependence.
2	Loss of soluble silica in the form of $Si(OH)_4$ to the solution, resulting from breaking of Si–O–Si bonds and formation of Si–OH (silanols) at the glass solution interface: $Si\text{-}O\text{-}Si + H_2O \rightarrow Si\text{-}OH + OH\text{-}Si$ This stage is usually controlled by interfacial reaction and exhibits a $t^{1.0}$ dependence.[60,61]
3	Condensation and repolymerization of a SiO_2-rich layer on the surface depleted in alkalis and alkaline-earth cations: $$\begin{array}{cccc} O & O & O & O \\ \| & \| & \| & \| \\ O\text{-}Si\text{-}OH + HO\text{-}Si\text{-}O \rightarrow O\text{-}Si\text{-}O\text{-}Si\text{-}O + H_2O \\ \| & \| & \| & \| \\ O & O & O & O \end{array}$$
4	Migration of Ca^{2+} and PO_4^{3-} groups to the surface through the SiO_2-rich layer forming a $CaO\text{-}P_2O_5$-rich film on top of the SiO_2-rich layer, followed by growth of the amorphous $CaO\text{-}P_2O_5$-rich film by incorporation of soluble calcium and phosphates from solution.
5	Crystallization of the amorphous $CaO\text{-}P_2O_5$ film by incorporation of OH^-, CO_3^{2-}, or F^- anions from solution to form a mixed hydroxyl, carbonate, fluorapatite layer.

When bioactive glasses are exposed to solutions, such as distilled water, or simulated body fluids or *in vivo*, the surface undergoes a series of chemical reactions, as summarized in Table 1. A result of the chemical reactions is the formation of a layer of hydroxy-carbonate apatite (HCA) crystallites which bond to various biological constituents, including collagen fibers. Bone growing cells, osteoblasts, attach to the growing HCA layer and eventually new bone is bonded across the interface. This series of interfacial reactions is summarized in Fig. 3. Note that the first five stages involve physical-chemical changes to the material which occur prior to the biochemical interactions, Stages 6-12. As discussed below, FTIR can be used to measure quantitatively the kinetics of Stages 1-5.

The reaction sequence depicted in Table 1 and Fig. 3 results in a compositional profile of the glass-solution interface that has two distinct layers, (1) a repolymerized SiO_2-rich layer on the glass and (2) a CaO-P_2O_5-rich layer between the glass and tissue. Auger electron spectroscopy combined with Ar-ion milling shows that these two layers develop as early as 1 hour after implantation (Fig. 4). Electron microprobe analyses of a bone-bonded Bioglass® interface after one year (Fig. 5) still has the two layers with SiO_2-rich layer growing to about 70 μm thick and the CaO-P_2O_5-rich layer being about 30 μm thick. Thus, it is critical to understand the kinetics of these layers which can be done with FTIR. Also quality control of bioactive glasses in either bulk form, powders, or coatings can be based on the formation of the two reaction layers using standard test conditions.

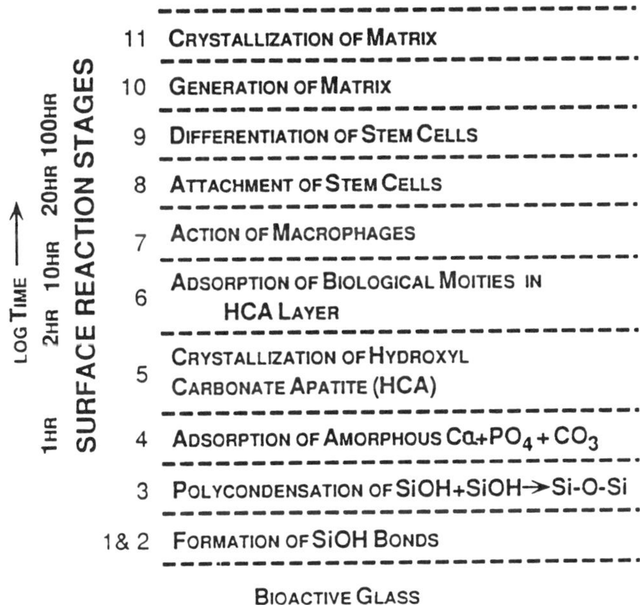

Figure 3. Sequence of interfacial reactions involved in forming a bond between tissue and bioactive ceramics.

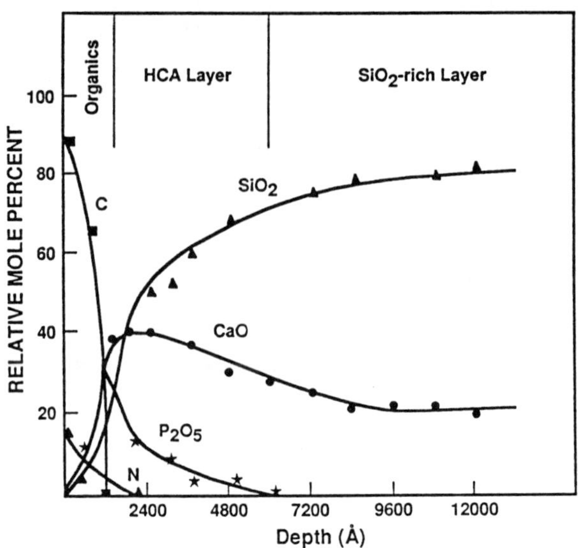

Figure 4. Bilayer films formed on 45S5 Bioglass® after 1 h in rat bone, *in vivo* (1 Å = 10⁻¹ nm)

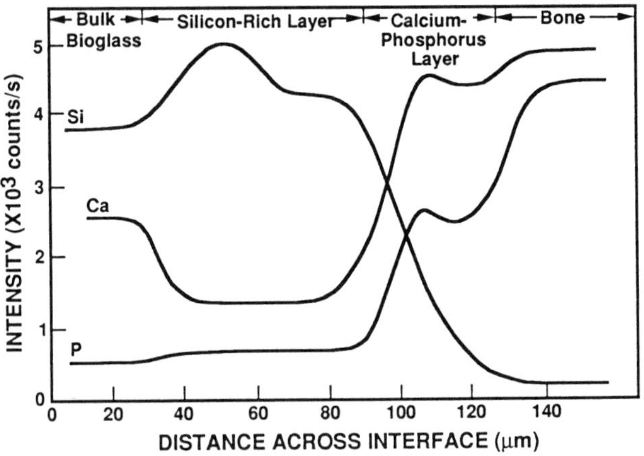

Figure 5. Electron microprobe analysis across an implant-bone interface. (Electron microprobe at 20 kV, 100 nA (specimen current), ~1 μm beam diameter, and 20 μm (min·in.) scan rate.)

     Characterization Techniques for the Solid-Solution Interface

Materials and Methods

The samples used for this study were prepared by the melt derived method or by the sol-gel process. The melt derived method was used to make 1 cm diameter disks and glass frit which was grounded and then passed through stainless steel sieves to establish the desired particle size ranges. The sol-gel processed powders were also sieved to the desired particle size ranges. The solid disks were dry polished with 320 and 600 grit SiC paper then sonicated in acetone for 15 minutes. All samples were reacted in 30 ml nalgene bottles for various times in Tris buffer solution with a pH of 7.4 at 37°C in a constant temperature bath. The solid disks were suspended in solution using fine polypropylene string and powders were placed in the bottom of the container. A surface area to volume ratio (SA/V) of 0.1 cm^{-1} was used to eliminate any volume effects. All samples were carefully rinsed in acetone for thirty seconds after removal from solution to stop any further reaction.

All IRRS analysis was completed using a Nicolet 20SXB spectrometer with a triglyceride sulfate (TGS) detector. The FTIR bench was purged with nitrogen to eliminate water and CO_2 vapors. All spectra were collected from 4000 to 400 wavenumbers at two wavenumber resolution and were ratioed against a front coated mirror to eliminate any instrumental response. The specular reflectance data were collected using a Perkin Elmer combination specular reflectance stage. The diffuse reflectance data was collected using a DRIFTS accessory from Spectra Tech, Inc.

SPECULAR REFLECTANCE

A schematic diagram for the specular reflectance stage used can be seen in Fig. 6. A series of flat front coated mirrors direct the IR radiation to the sample platform. The spot size for the analysis is determined by the aperture on which the sample rests. For our analysis a spot size of 0.5 cm was used. After reflection from the sample surface a series of flat mirrors were directed to reflect radiation to the detector. Because this system uses flat mirrors and an angle of incidence of approximately 30° from the normal, a strong specular component results with little or no diffuse component measured.

The FTIR spectra of a 45S5 Bioglass® sample before (0 hrs) and after 2 hours and 20 hrs reaction in a Tris buffer solution (pH 7.4) are shown in Fig. 7. These spectra were collected using 128 scans which takes about five minutes. The primary Si-O stretching (1100 cm^{-1}) and rocking (450 cm^{-1}) vibrational modes are present in the glass before reaction along with the Si-O-Na, Ca stretching vibrations at 800-925 cm^{-1}. Alkali ion exchange (Stage 1), network dissolution (Stage 2) and silica condensation reactions (Stage 3) produce a surface spectrum at 2 hrs that is distinctly different from the 0 hr spectrum. However, specular reflection does not indicate formation of the HCA layer (Stages 4 and 5) within 2 hours. At 20 hours, however, the HCA layer is distinct with the three vibrational modes characteristic of polycrystalline HCA visible at 500 cm^{-1}, 525 cm^{-1}, and 575 cm^{-1}. The quality of these spectra are fairly good but one can see that the noise level increases as the surface becomes more irregular with reaction time.

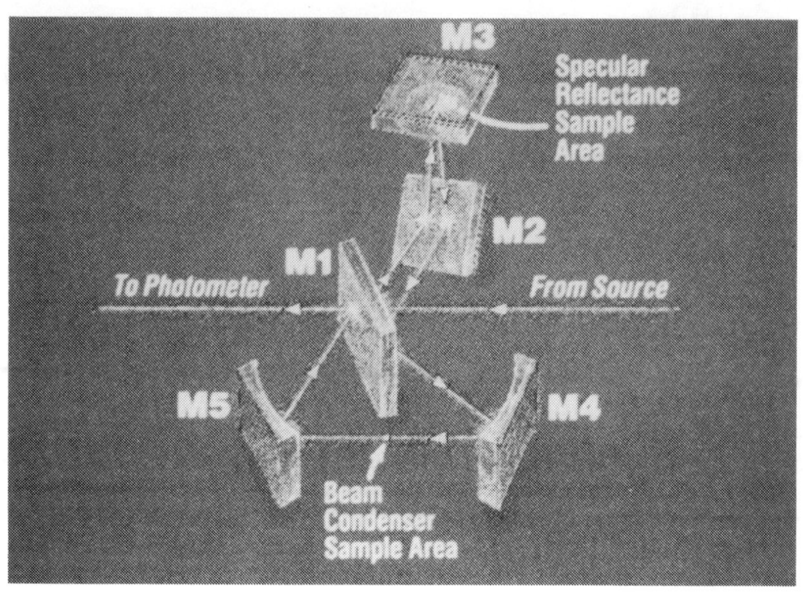

Figure 6. Specular reflectance stage schematic.

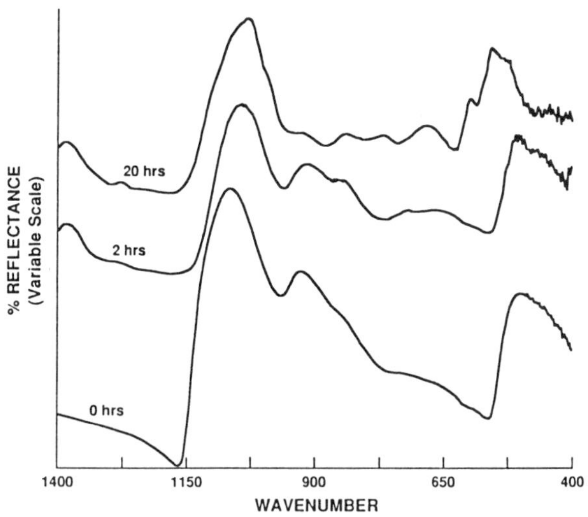

Figure 7. Specular Reflectance spectra of 45S5 reacted in Tris buffer.

    Characterization Techniques for the Solid-Solution Interface

No specular reflection data is shown for Bioglass® powders since the technique is not suitable due to excessive scattering from the powder surface and difficulties with the sample mount of powders.

DIFFUSE REFLECTANCE

A schematic diagram for the diffuse reflectance stage used is seen in Fig. 8. A series of flat front coated mirrors direct the IR radiation to a parabolic mirror which focuses the incident beam onto the sample surface. The sample stage can accommodate almost any kind of solid sample including smooth, flat solids and powders.

After reflection from the sample surface all infrared radiation is collected on the other side of the parabolic mirror and focused onto a series of flat mirrors which direct the reflected radiation to the detector. By use of the parabolic mirror the diffuse stage is able to collect both specular and diffuse components of the reflected beam. This again makes possible the analysis of a variety of different sample configurations.

The diffuse reflectance method is substantially more sensitive to solution reactions at a glass interface. Figure 9 shows the 0 hr, 1 hr, and 2 hr spectra of a 45S5 glass, identical to the specimen shown in Fig. 7. These spectra were collected using 32 scans which takes about 1 minute. Spectral features before reaction are equivalent, when compared to the specular results, as expected. However, in just one hour the FTIR diffuse reflection spectrum shows the onset of Stage 4, precipitation of an amorphous $CaO\text{-}P_2O_5$-rich layer, on top of the SiO_2-rich layer. The intensity of the P-O vibrations is low but the peak at 600 cm^{-1} is quite visible and grows with time, as discussed later. This increase in sensitivity can be attributed to the ability of the diffuse stage to collect any reflected radiation including surface scattering and internal or attenuated energy from the sample surface, unlike the specular reflectance stage.

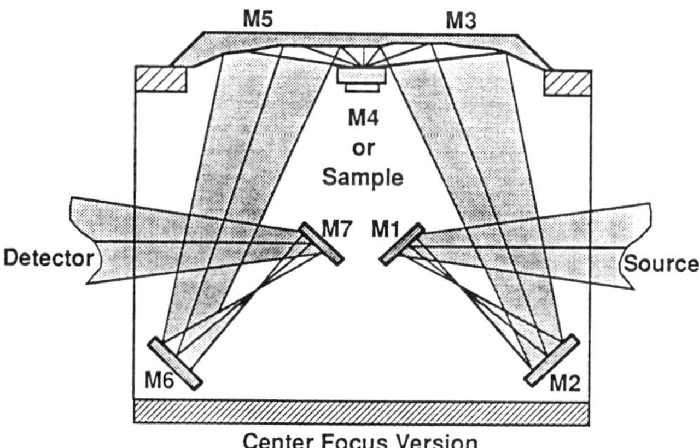

Figure 8. Schematic diagram of diffuse reflectance stage from Spectra Tech Inc.

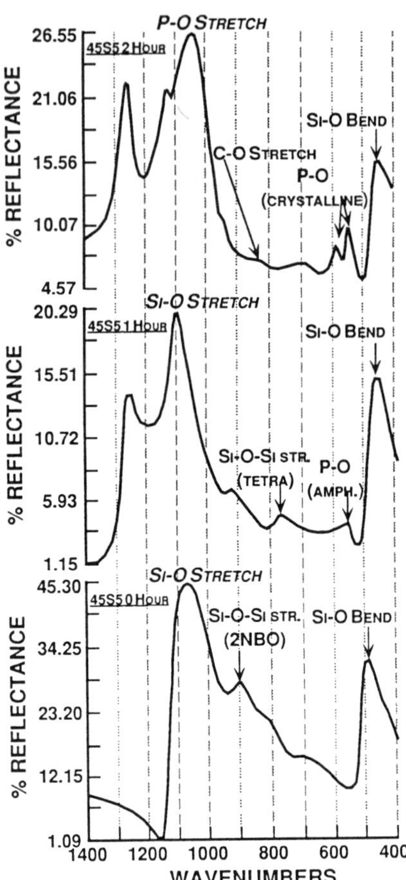

Figure 9. FTIR spectra of a 45S5 Bioglass® reacted for 0, 1, and 2 h in Tris buffer at 37°C.

Within two hours the diffuse reflection spectrum shows that the glass surface has developed a polycrystalline HCA layer with both P-O vibrations and C-O vibrations present in the new crystalline phase. Comparison of Figs. 7 and 9 show again that the diffuse reflection method is extraordinarily more sensitive than specular reflection to analyze the resultant surface composition as a result of a solid-solution interaction.

Even Bioglass® powders can be analyzed with the diffuse reflection method. Figure 10 shows the spectra of three different particle size ranges used in repair of periodontal defects and bone augmentation. The spectrum before reaction is compared with a spectrum of a bulk 45S5 Bioglass®. Except for a decrease in intensities all spectral features of the bulk surface are present for the powders (Fig. 10). A reaction sequence of 45S5 Bioglass® powders from 2 minutes to 120 minutes in Tris buffer (pH 7.4) is shown in Fig. 11. Comparison of Fig. 11 with Fig. 9 shows that the powders follow

   Characterization Techniques for the Solid-Solution Interface

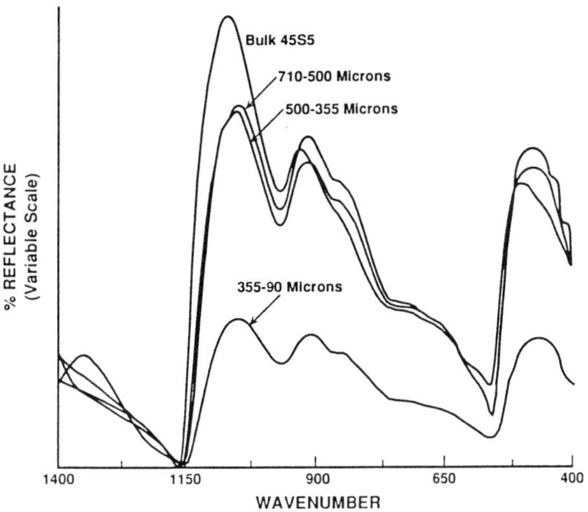

Figure 10. The effects of particle size on the diffuse spectra of 45S5 powders.

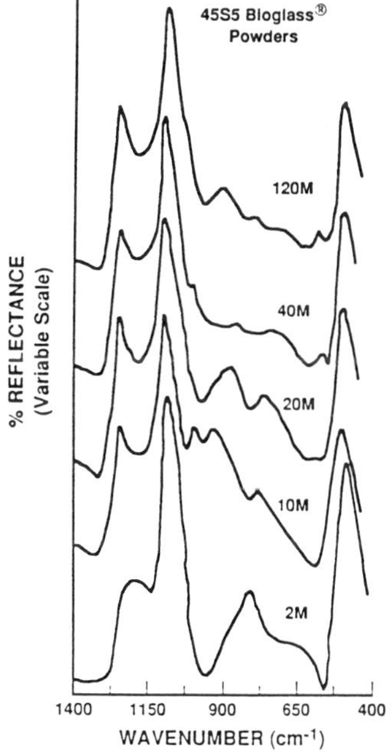

Figure 11. FTIR spectra for 45S5 Bioglass® powder with various reaction time at early stages.

Characterization Techniques for the Solid-Solution Interface

187

the same reaction sequence at about the same rate as does the bulk glass surface. Thus, the FTIR diffuse reflection method is suitable for QC/QA tests of either bulk <u>or</u> powdered glasses. Warren discusses the relative merits of using static test solution versus stirred, dynamic solutions with Bioglass® powders followed by FTIR diffuse reflection spectroscopy.[26] A standard QA/QC test for both Bioglass® powders and bulk implants is based upon these results.

IR MICROSCOPY

A schematic diagram of an IR microscope can be seen in Fig. 12. The technique combines the use of optical microscopy with infrared spectroscopy. The IR microscope enables the user to optically focus the infrared beam down to an approximately 10 micron spot size. By use of advanced Cassegrain optics the system can be run in either reflection or transmission modes.

The IR microscope provides a fast, inexpensive means for analyzing heterogeneities or establishing the homogeneity of a material. An example is shown in Fig. 13. Eight spots of 10 μm diameter were analyzed following a 20 hr QC test of a 45S5 Bioglass® sample in 37°C Tris buffer (7.4 pH) solution. The location of the primary P-O stretching vibration at 1050 cm^{-1} is consistent across the sample, as illustrated in the insert. This establishes the uniformity of the HCA reaction layer that grows on the glass.

The IR microscope can also be used to measure compositional profiles across a sample, as shown in Fig. 14. The silica-rich and HCA layers are clearly distinguishable in the spectra from a cross section of the Bioglass® sample reacted for 72 hours in 37°C Tris buffer. These results are consistent with the profiles obtained by electron beam

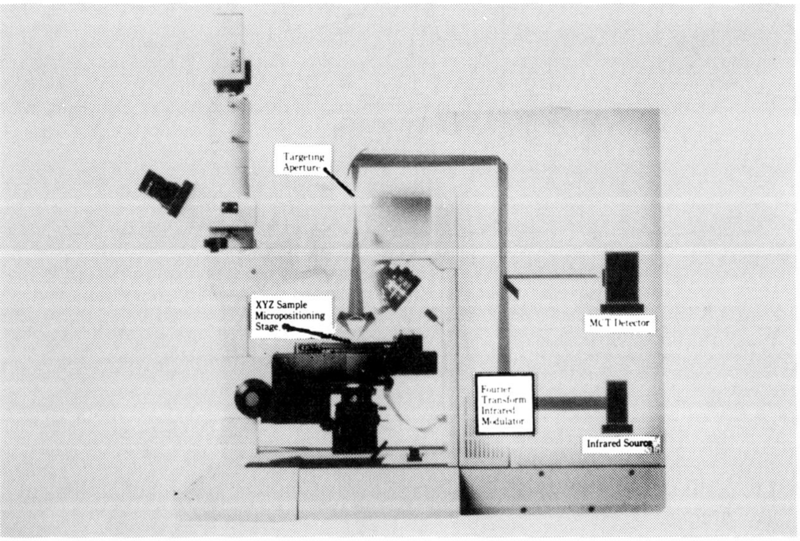

Figure 12. Schematic diagram of the infrared microscope in the reflection mode.

      Characterization Techniques for the Solid-Solution Interface

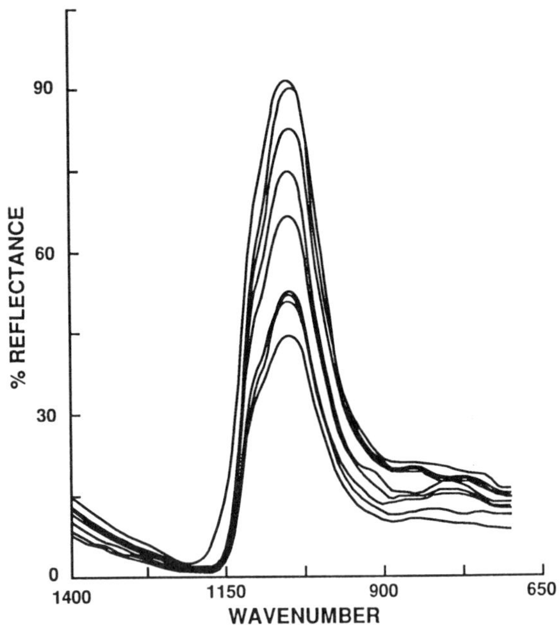

Figure 13. IR microscope profile across the surface of a 1 cm 45S5 disc for 20 hrs. in Tris buffer.

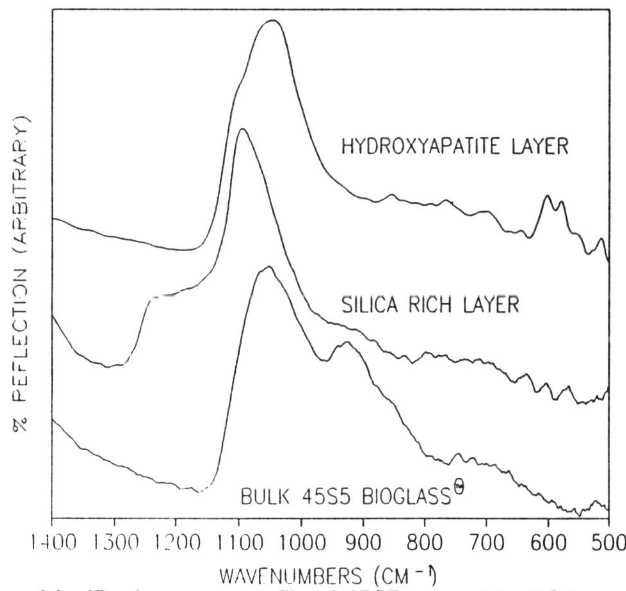

Figure 14. IR microscope profile of 45S5 reacted for 72 hours.

Characterization Techniques for the Solid-Solution Interface

techniques shown earlier (Figs. 4 and 5). However, the IR microscope offers the advantage of determining whether the calcium-phosphate-rich layer is amorphous or crystalline whereas E-beam methods yield only compositional data.

An important use of the IR microscope is analysis of the components of heterogeneous structures. An example of great complexity is the analysis of the repair of a bone defect packed with a mixture of Bioglass® powder and autogeneous bone chips from the same animal. The mixture of Bioglass® and bone produces a volume of bone that is 12 times the volume of the original bone (dog rib bone). However, the repairing structure is a mixture of old bone, new cartilage, new bone, partially mineralizing cartilage, soft tissue, and the remnants of the glass powder. Figure 15 shows the spectra of two spots from this heterogeneous material. One spectrum is characteristic of bone mineral with P-O vibrational modes. The other spectrum is characteristic of a highly alkali-depleted silica-rich phase, a relic of the original Bioglass® particle. After 7 weeks implantation, the duration of the experiment, the small glass particle has reacted throughout its volume and is no longer physically distinguishable as glass except for the IR vibrational modes.

APPLICATIONS

One of the most important uses of FTIR analysis of Bioglass®-solution interactions is to measure the kinetics of Stages 1-5 reactions, shown in Table 1 and Fig. 3. Plots of the time dependent changes in the location of the various surface vibrational modes create a kinetic profile of the material. Figure 16 is such a plot for 45S5 Bioglass® in bulk form reacted in Tris buffer at 37°C. The time for onset of each of the five reaction stages is measured with an accuracy of ±5 minutes. The onset of HCA crystallization from the amorphous calcium-phosphate layer is clearly distinguished, as

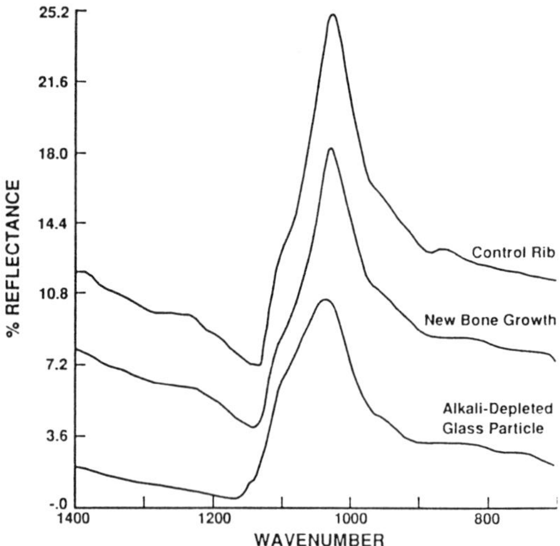

Figure 15. IR microscope spectra from a cross section of a dog rib augmented with Bioglass® powder.

    Characterization Techniques for the Solid-Solution Interface

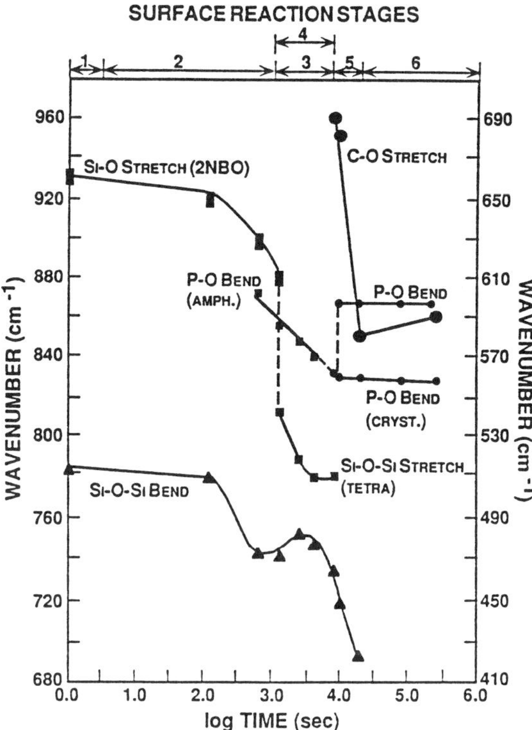

Figure 16. Time-dependent changes in IR vibrations of the surface of 45S5 Bioglass®
reacted in a 37°C Tris buffer solution.

is the incorporation of CO_2 within the structure to form the crystalline phase. The decreasing vibrational frequency of the Si-O modes prior to complexation of Ca-P on the surface indicates nucleation of the new phase by the silanol condensation reaction.

Another important application of the FTIR kinetics analysis is in the development of new bioactive materials. Sol-gel processing by Li, et al.[27] has recently led to the production of a new generation of alkali-free bioactive gel-glasses. A wide compositional range of the SiO_2-CaO-P_2O_5 gel-glasses develop a HCA layer and are therefore bioactive. Figure 17 shows FTIR results after just 1 hour in Tris-buffer test solution from compositions containing 49 weight % SiO_2 (49S) to 77 weight % SiO_2 (77S). See reference 27 for details of the compositions. Comparison of the FTIR results from Fig. 17 with Fig. 11 shows that the gel-glasses develop a crystalline HCA layer more rapidly than melt derived glasses even with more SiO_2 in the structure. The compositional limit for bioactivity is 60% SiO_2 for melt glasses whereas it has been extended to approximately 90% SiO_2 for the gel-glasses. As shown by Li, et al.,[27] FTIR was the critical analytical method in the development of these new materials.

Characterization Techniques for the Solid-Solution Interface   191

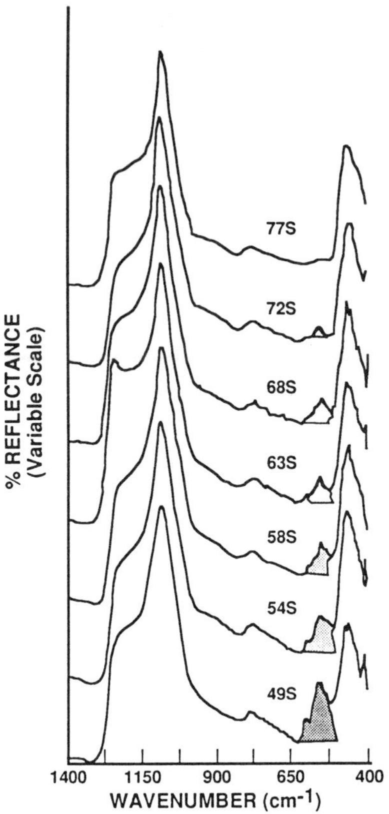

Figure 17. FT-IRRS spectra of gel powders with various composition after 1 h.

SUMMARY

Fourier transform IR spectroscopy is a versatile and inexpensive method for analyzing glasses, glass-ceramics, and ceramics as bulk materials, coatings, or powders. Interfacial reactions that produce chemical changes in the surface are easily followed by FTIR, especially using a diffuse reflection stage. Heterogeneities and profiles of interfacial reaction zones due to solution reactions can be analyzed using the IR microscope on an FTIR. Spot sizes as small as 10 μm can be analyzed.

ACKNOWLEDGMENTS

The authors are grateful for the financial support of the Air Force Office of Scientific Research under contract #F49620-88-C-0073.

REFERENCES

1. D. M. Sanders, W. B. Person and L. L. Hench, "New Methods for Studying Glass Corrosion Kinetics," Appl. Spectroscopy 26[5] (1972) 530-536.

2. D. M. Sanders and L. L. Hench, "Mechanisms of Glass Corrosion," J. Am. Ceram. Soc. 56[7] (1973) 373-377.

3. D. M. Sanders and L. L. Hench, "Environmental Effects on Glass Corrosion Kinetics," Bull. Am. Ceram. Soc. 52[9] (1973) 662-665.

4. D. M. Sanders and L. L. Hench, "Surface Roughness and Glass Corrosion Kinetics," Bull. Am. Ceram. Soc. 52[9] (1973) 666-669.

5. D. M. Sanders, W. B. Person and L. L. Hench, "Quantitative Analysis of Glass Structure Using Infrared Reflection Spectra," Appl. Spectroscopy 28[3] (1974) 247-255.

6. L. L. Hench and D. M. Sanders, "Analysis of Glass Corrosion," Glass Industry, (1974), 13.

7. L. L. Hench, "Characterization of Glass Corrosion and Durability," J. Non-Crystalline Solids 19 (1975) 27-39.

8. D. E. Clark, M. F. Dilmore, E. C. Ethridge and L. L. Hench, "Aqueous Corrosion Behavior of Soda-Lime Silica and Soda-Lime-Silica Glass," J. Am. Ceram. Soc. 59[1-2] (1976) 62-65.

9. D. E. Clark, E. C. Ethridge, M. F. Dilmore and L. L. Hench, "Quantitative Analysis of Corroded Glass Using IRRS Frequency Shifts," Glass Technol. 18[4] (1977) 121-124.

10. M. F. Dilmore, D. E. Clark and L. L. Hench, "Chemical Durability of a Na_2O-K_2O-CaO-SiO_2 Glass," Bull. Am. Ceram. Soc. 61[9-10] (1978) 439-443.

11. D. E. Clark, E. C. Ethridge and L. L. Hench, "Effects of Glass Surface Area to Solution Volume Ratio on Glass Corrosion," Physics and Chemistry of Glasses 20[2] (1979) 35-40.

12. M. F. Dilmore, D. E. Clark and L. L. Hench, "Aqueous Corrosion of Lithia-Alumina-Silicate Glasses," Bull. Am. Ceram. Soc. 57 (1978) 1040-1044.

13. M. F. Dilmore, D. E. Clark and L. L. Hench, "Corrosion Behavior of Lithia Disilicate Glass in Aqueous Solutions of Aluminum Compounds," Am. Ceram. Soc. Bull. 58[11] (1979) 1111-1114.

14. C. G. Pantano, Jr., D. E. Clark, and L. L. Hench, Corrosion of Glass, The Glass Industry, New York, 1979.

15.   L. L. Hench and D. E. Clark, "Physical Chemistry of Glass Surfaces," J. Non-Crystalline Solids 28[1] (1978) 83-105.

16.   L. L. Hench, "Corrosion of Silicate Glasses: An Overview," in _Materials Stability and Environmental Degradation_, Aaron Barkatt, Ellis D. Verink, Jr., and Leslie R. Smith, eds., Vol. 125, Materials Research Society, Pittsburgh, PA, 1988, pp. 189-200.

17.   L. L. Hench, D. E. Clark and A. B. Harker, "Nuclear Waste Solids," Journal of Materials Science 21 (1986) 1457-1478.

18.   Larry L. Hench, "International Collaboration in Nuclear Waste Solidification," Nuclear Technology 73 (1986) 188-198.

19.   L. L. Hench, R. J. Splinter, T. K. Greenlee and W. C. Allen, "Bonding Mechanisms at the Interface of Ceramic Prosthetic Materials," J. Biomedical Materials Research, (November 1971), No. 2, Part 1, 117-141.

20.   L. L. Hench, "Biomedical Applications and Glass Corrosion," Proceedings of Xth Intl. Congress on Glass, Ceramic Society of Japan, Kyoto, Japan (1974) 30-41.

21.   L. L. Hench, "Stability of Ceramics in the Physiological Environment," _Fundamental Aspects of Biocompatiblity_, D. F. Williams, ed., CRC Press, Boca Raton, Florida, Vol. I, 1981, pp. 67-85.

22.   Cheol Y. Kim, A. E. Clark and L. L. Hench, "Early Stages of Calcium-Phosphate Layer Formation in Bioglass," J. Non-Crystalline Solids 113 (1989) 195-202.

23.   L. L. Hench, O. A. Andersson and G. P. LaTorre, "The Kinetics of Bioactive Ceramics, Part III: Surface Reactions for Bioactive Glasses Compared with an Inactive Glass," in _Bioceramics, Volume 4_, W. Bonfield, G. W. Hastings and K. E. Tanner, eds., Butterworth-Heinemann Ltd., Guildford, England, 1991, pp. 155-162.

24.   L. L. Hench, "Surface Reaction Kinetics and Adsorption of Biological Moieties: A Mechanistic Approach to Tissue Attachment," in _The Bone-Biomaterial Interface_, J. E. Davies, ed., University of Toronto Press, 1991, pp. 33-44.

25.   L. L. Hench, "Bioactive Ceramics," in _Bioceramics: Materials Characteristics Versus In Vivo Behavior_, Paul Ducheyne and Jack E. Lemmons, eds., Annuals of the New York Academy of Sciences, Vol. 523, 1988, pp. 54-71.

26.   L. Warren, "Investigation of Bioglass® Powders and Pastes," M.S. Thesis, University of Florida, 1987.

27.   R. Li, A. E. Clark, and L. L. Hench, "Bioactive Glass Powders," J. Appl. Biomaterials 2 (1991) 231-239.

   Characterization Techniques for the Solid-Solution Interface

28.   D. A. Skoog and D. M. West, <u>Principles of Instrumental Analysis</u>, 2nd edition, Saunders College, 1980, pp. 209-254.

29.   James E. Stewart, <u>Infrared Spectroscopy, Experimental Methods and Techniques</u>, Marcel Dekker, Inc., New York, 1970.

30.   E. G. Brame, Jr. and J. G. Grasselli, editors, <u>Infrared and Raman Spectroscopy</u>, Part A, Marcel Dekker, Inc., New York, 1982.

31.   Peter R. Griffiths and Michael P. Fuller, "Infrared Spectroscopy of Powdered Samples," in <u>Advances of Infrared and Raman Spectroscopy</u>, Vol. 9., R.J.H. Clark and R. E. Hesto, eds., Heyden and Sons, Ltd., Philadelphia, PA, 1982, pp. 63-129.

32.   K. Lawson, <u>Infrared Absorption of Inorganic Substances</u>, Reinhold, New York, 1961.

Characterization Techniques for the Solid-Solution Interface   195

CHARACTERIZATION OF SURFACTANT AND POLYMER AGGREGATES AT THE CERAMIC SOLID - SOLUTION INTERFACE USING IN - SITU SPECTROSCOPIC TECHNIQUES

P. Somasundaran and C. Maltesh
Langmuir Center for Colloids & Interfaces
Columbia University, New York, NY 10027.

INTRODUCTION

There has been an increasing awareness recently of the value of applying the principles of surface and colloid science to ceramics processing. Colloidal processing techniques offer greater control over mixing and packing density variations as compared to dry powder routes. Particles in a colloidal suspension can be dispersed through particle-particle repulsive forces resulting from electrostatic interaction, steric hindrance, or a combination of both[1]. Such particle-particle repulsion helps assure both the breakup of the soft agglomerates held together by van der Waal's forces and the formation of well-dispersed suspensions, even in multi-phase systems. Colloidal suspensions may be effectively used to eliminate unwanted flaw origins, such as hard agglomerates and particles larger than a certain size[2]. It has thus become the major objective of ceramic processing engineers to achieve a better control of the state of the dispersion of the powder processed and of the consequent suspension stability[3]. There has been considerable work done recently on techniques to control the stability of a dispersion through judicious adsorption of polymers and surfactants. Most of these are based on an empirical approach primarily due to lack of suitable techniques to monitor the processes *in-situ* and on a molecular scale. Extensive work has been done on understanding the physico-chemical interactions that govern the formation of adsorption layers using methods involving determination of adsorption isotherms, zeta potential, hydrophobicity and heat of adsorption. While these methods provide useful information on a mechanistic level, the importance of structure of the adsorbed layers in controlling the interfacial properties has been noted recently[4]. For example, in addition to the extent of surfactant adsorption on particles, the orientation of the adsorbed surfactant molecules will indeed influence their flotation. Similarly, conformation of the polymeric species play a major role in the flocculation/dispersion of colloidal particles and information on the micro and nanostructure of the adsorbed layers can help to manipulate the system behavior.

Spectroscopic techniques are suitable for generating information on a molecular scale at the solid-liquid interface without disturbing the equilibrium. Spectroscopy requires the presence of intrinsic or extrinsic probes. Spectroscopic methods based on luminescence emission and paramagnetic resonance require an appropriate spectroscopically sensitive label whereas techniques like nuclear magnetic resonance (NMR), infra-red (IR) and Raman utilize the inherent nuclear spin and vibrational modes. When using an externally added probe, care should be taken to ensure that the probe itself is inert and does not perturb the equilibrium and dynamics of the process. Recently, research in this laboratory has been directed towards developing spectroscopic techniques (fluorescence spectroscopy, electron spin resonance spectroscopy (ESR), and time resolved Raman spectroscopy) for examining the adsorbed layers at the solid-liquid interface in-situ. The use of these techniques is based on the fact that the spectral responses of the probes are highly environment dependent and as such serve to characterize the environment in which they reside. Various probes used in these studies are depicted in figure 1.

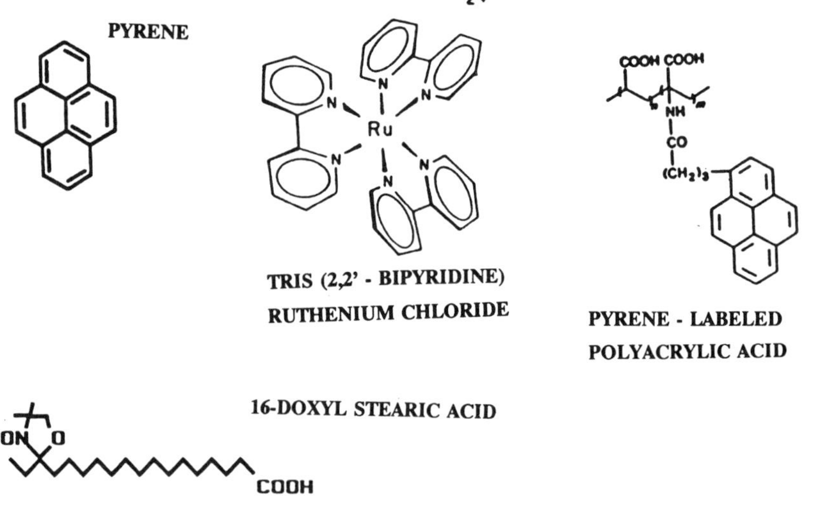

Figure 1: Various probes used in the spectroscopic studies

Principles of the different techniques and the information obtained from them are discussed in the following sections.

PRINCIPLES OF IN-SITU SPECTROSCOPIC TECHNIQUES

Fluorescence Spectroscopy[5]

Certain organic molecules in their ground state P when excited by light, absorb incident light energy and reach an electronically excited state P^*. The transition of the molecule from the excited state P^* to the ground state P is accompanied by the emission of light which is referred to as luminescence. Complex organic molecules that are commonly used as luminescence probes typically absorb energy in the spectral range 250 to 650 nm which corresponds to transitions with energy changes of 2-5 eV. Several features of luminescence, such as fluorescence, phosphorescence, excimer fluorescence and delayed fluorescence can be used as tools to obtain information at the molecular level on polymers and surfactants in the bulk and at the solid-liquid interface. The luminescence experiment essentially involves measurement of changes in the emission properties of a probe or its photochemical intermediates in order to determine the nature and rates of photophysical and photochemical processes in the system. In this respect, fluorescence is the most widely used technique. Fluorescence responses are sensitive to changes in the microenvironment around the probe so that a luminescence probe in different microenvironments will display experimentally distinct luminescence properties characteristic of each microenvironment. The fluorescent probe utilized here is pyrene. A typical emission spectrum of pyrene is shown in Figure 2.

The excimer formation tendency of pyrene is widely exploited for analysis of aggregation[6] and conformational studies[7]. An excimer is a dimer formed between an excited molecule and a molecule in the ground state. The photophysics of pyrene monomer (eq. 1) and excimer formation (eq. 2) can be represented as follows:

$$P + h\nu \rightarrow P^* \rightarrow P + h\nu' \tag{1}$$

$$P + P^* \rightarrow (PP)^*_{excimer} \rightarrow P + P + h\nu'' \tag{2}$$

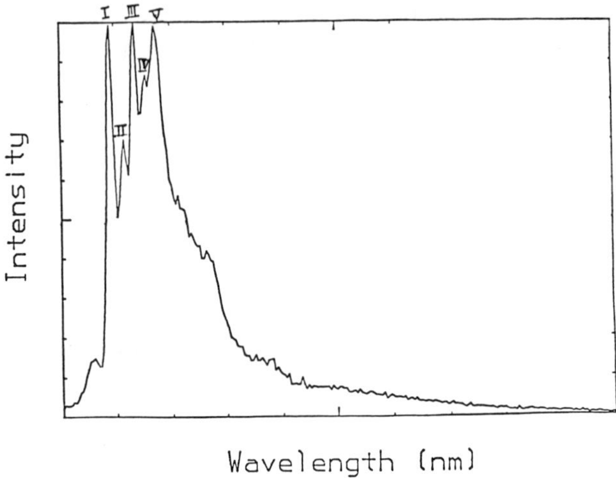

Figure 2: A typical emission spectrum of pyrene

Excimer fluorescence in pyrene is observed as a broad spectral band centered around 480 nm that is red shifted with respect to the monomer emission. The excimer complex is formed only when the aromatic rings approach each other within 0.4-0.5 nm[8].

The extent of excimer formation will depend upon a number of factors, viz., number of probes in the system and the viscosity of the environment in which the probe resides. In a fragmented media such as surfactant micelles or surfactant aggregates at the solid-liquid interface, the distribution of the probe molecules throughout the system as well as aggregate size will affect the observed decay profiles[9]. In micellar media it is generally assumed that the probes are distributed among the micelles according to Poisson statistics. It has been shown that such assumptions are valid. Based on this and on the photophysics of pyrene fluorescence, the decay in fluorescence intensity with time is given by:

$$I_t = I_0 \exp\left[-k_m \cdot t - n\left(1 - \exp\{k_e \cdot t\}\right)\right] \qquad (3)$$

where n is the average number of probes per aggregate, k_m is the rate constant of decay of an excited pyrene monomer, and k_e is the excimer formation-dissociation rate constant. At long times, the fluorescence decay profiles represent the decay due to monomeric emission in the absence of

                Characterization Techniques for the Solid-Solution Interface

excimer formation. Therefore equation 3 reduces to:

$$\ln\left[\frac{I_t}{I_0}\right] = -n - k_o \cdot t \qquad (4)$$

and extrapolation to t=0 gives n, the average number of probe in each aggregate. For micellar systems (or for that matter, any hydrophobic aggregate), n is defined as:

$$n = \frac{[P]}{[Agg]} = \frac{[P] \cdot N}{C_o - CMC} \qquad (5)$$

where, [P] is the probe concentration and C_o is the total surfactant concentration. For surfactant aggregates at the solid-liquid interface, instead of the critical micellization concentration (CMC) the equilibrium or residual concentration can be used.

Another application of excimer fluorescence of pyrene is in determining polymer conformation. Pyrene can be randomly attached onto a polymer chain ensuring that the amount of pyrene is low so that it does not affect the polymer dissolution characteristics significantly[10]. Since the pyrene is covalently bonded to the polymer chain, excimer formation will depend upon the conformation of the polymer. Depending upon the solution conditions, if the polymer is coiled we would detect significant excimer formation. If on the other hand the polymer were stretched or extended then the probability of two pyrene molecules coming close to one another would be low and subsequently excimer formation would be low. The ratio of excimer and monomer intensities would then provide a measure of the conformation of the polymer. A high value of this ratio would indicate a coiled polymer and a low value would indicate a stretched polymer. The ratio can therefore be termed as the *coiling index*. It must be noted here that excimer formation would be a true indicator of polymer conformation only if the excimer formation were intramolecular. Intermolecular excimer formation can be avoided by maintaining dilute solutions of the labeled polymer. If concentrated solutions need to be studied a mixture of labeled and unlabeled polymers should be used. A simple test to differentiate between inter- and intramolecular excimer formation is to determine the value of the coiling index as a function of polymer concentration. If the excimer formation is intramolecular then the ratio would not be affected by polymer concentration, however if it were intermolecular it would increase with increase in polymer concentration.

Electron Spin Resonance Spectroscopy[11]

Electron Spin Resonance (ESR) is a form of spectroscopy that is well suited to the study of molecular structure and dynamics. It is a technique dependent on transitions between spin levels of molecular unpaired (paramagnetic) electrons in an external magnetic field in the form of absorption of microwave radiation. The intrinsic angular momentum of a free electron splits in an external magnetic field and undergoes hyperfine splitting upon the influence of secondary magnetic moments of neighboring nuclei. The range of applications of ESR can be extended by spin labeling methods where stable free radicals are incorporated into systems of interest so as to characterize their dynamic and physical properties. The nitroxide radical is commonly used in spin labeling. In this study, doxyl stearic acid (stearic acid labeled with a nitroxide bearing moiety) was chosen as the probe owing to the fact that this structure resembles the dodecylsulfate molecules to some extent. The anionic functionality and long alkyl chain should enable the probes to coadsorb with the sodium dodecylsulfate. A typical spectrum of a doxyl stearic acid is shown in figure 3.

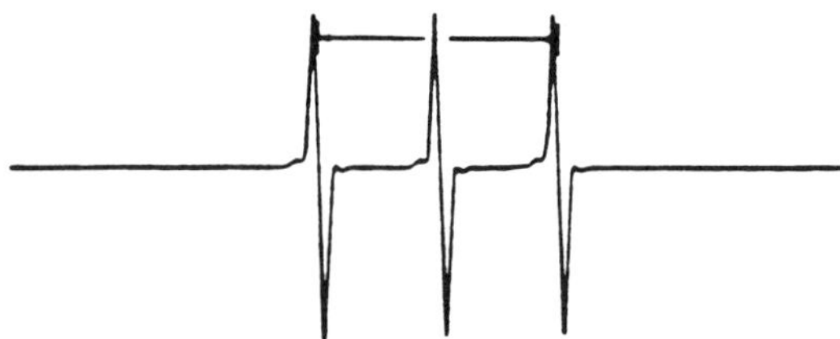

Figure 3: ESR spectrum of nitroxyl radical labeled stearic acid

Several pieces of information can be obtained from the ESR response of these type of probes[12]:

1. Polarity of the probe environment: this can be used to reveal the location of the probe.
2. Viscosity/structural ordering of the probe environment: hindrance in the rotational motion of the probe induces line broadening of the spectrum which can be used to access the fluidity of the probe environment.

          Characterization Techniques for the Solid-Solution Interface

3. State of aggregation of the probe molecules: when two probe molecules interact, the resulting spectrum is characteristic of a phenomenon called spin-spin relaxation. This feature can be exploited to describe the state of aggregation of molecules at the solid-liquid interface.

Resonance Raman Spectroscopy[13]

Raman spectroscopy gives information on the vibrational energy levels of molecules. The frequency shifts and the intensity changes of the Raman lines of the probe molecule can be utilized to characterize the environment of the probe. Raman spectroscopy is highly suitable for use in an aqueous environment compared to IR, due to the near transparency of the former and the ease with which the whole vibrational region of interest can be covered. While vibrational frequencies of molecules in the bound state should be different from that in the free state, they are also susceptible to changes in the symmetry properties of the environment. Very few definitive studies exist in the literature on the Raman spectroscopy of surfactants in solution. Raman investigations of surfactant adsorbates are reported as surface-enhanced Raman studies at the ceramic solid-solution interface.

The Raman probe used by us was tris (2,2' - bipyridine) ruthenium (II) chloride - $Ru(bpy)_3^{2+}$. The third harmonic of a Nd-YAG laser was the excitation source (pulse energy, 5 mJ; pulse width, 6 ns; wavelength; 354.5 nm). The spectrum of the probe was calculated using published spectra in literature. Figure 4 depicts the ground, excited and emitted state absorption spectra of the ruthenium complex. Figure 5 shows the ground state and excited state spectra of $Ru(bpy)_3^{2+}$ in water. The excited state consists of 14 lines, two sets of 7 lines each corresponding to the ground and excited state transitions. All these transitions are known to originate from the excited state species. They are assigned roughly to the various symmetric stretching vibrations of the C-N and C-C bonds of the bipyridine ring.

Infra-Red Spectroscopy

Infrared (IR) spectroscopy is concerned with light of wavelength ranging from 2000 to 5×10^7 nm. When infrared radiation is incident on an organic molecule, it is absorbed and converted to energy of molecular vibration. Molecular bonds between atoms are constantly undergoing stretching and bending motions at frequencies which depend upon the masses of the atoms involved and the type of chemical bond joining the atoms.

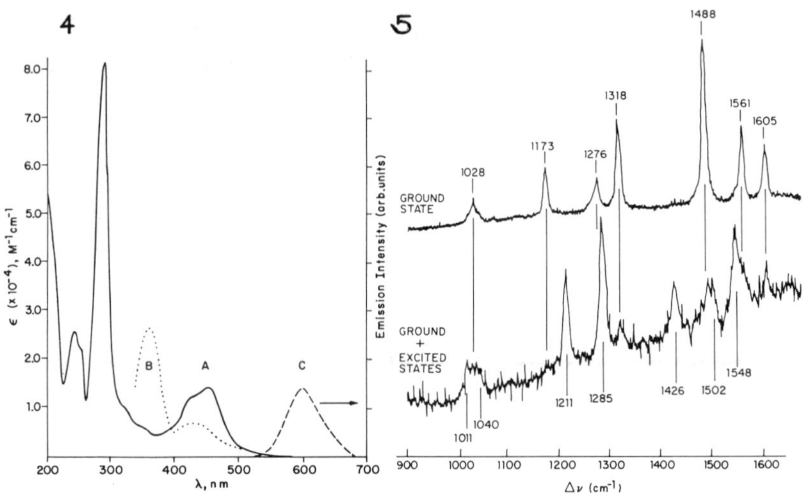

Figure 4:   **Absorption (A) and emission (C) spectra of Ru(bpy)$_3^{2+}$ and the excited state transient absorption spectrum (B) of Ru(bpy)$_3^{2+}$**

Figure 5:   **Raman spectra of ground and excited state Ru(bpy)$_3^{2+}$**

Since the frequencies of the various vibrations of the molecule correspond to those of IR radiation, absorption of the radiation occurs, producing an increase in the amplitude of the molecular vibrational modes. No irreversible change in the molecule results because the energy gained by the molecule in the form of light is soon lost in the form of heat. By plotting the ratio of light passing through the sample, I, to the intensity of light striking the sample, I_o, versus the frequency of the radiation, an IR spectrum is obtained. Instead of using incident radiation frequency, it is customary to use wavenumbers (units of cm^{-1}) which is the frequency, v, divided by the velocity of light, c, expressed in cm/sec. In Fourier Transform Infrared (FTIR) spectroscopy, which is more commonly used, the signal intensity measured at the detector is the cosine fourier transform of a wavenumber dependent spectral intensity. For additional information, there exist a host of excellent references in literature[14]. In practical applications, FTIR has been developed for solid and liquid samples. Depending on the absorption spectra, the mode of interaction between the adsorbate and adsorbent can be inferred. There are several variants of this technique such as diffuse reflectance infrared Fourier transform (DRIFT) and attenuated total reflectance (ATR) which have been designed for use with different kinds of samples.

 Characterization Techniques for the Solid-Solution Interface

CHARACTERIZATION OF ADSORBED SURFACTANT LAYERS

Surfactant Aggregation at the Solid-Liquid Interface: Effect on Dispersion Stability

It was realized in the early fifties that alkyl compounds could be used to improve the colloidal stability of carbon black-hydrocarbon suspensions without affecting the system's electrostatics, and that the stability can be optimized by judicious choice of chain length[15]. Subsequently, research conducted by one of the authors indicated similar behavior for anionic surfactants - oxide minerals in aqueous as well as nonaqueous media.[16,17,18] A model mineral/surfactant system studied in aqueous media was alumina/sodium dodecylsulfate. A typical adsorption isotherm for this system is shown in Figure 6. This isotherm is characterized by four distinct regions[19]: Region I, dominated by electrostatic adsorption; region II, marked by a sharp rise in the adsorption caused by surfactant aggregation on the particles; region III, characterized by a decreasing slope even though surfactant adsorption continues to increase and region IV, representing maximum surface coverage and marked by micelle formation in the bulk.

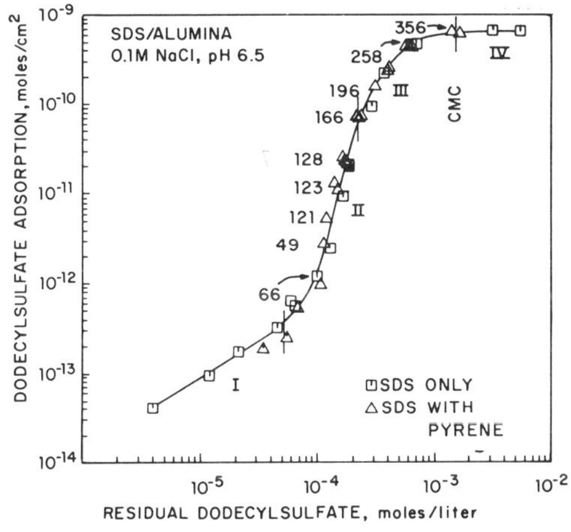

RESIDUAL DODECYLSULFATE, moles/liter

DODECYLSULFATE ADSORPTION, moles/cm²

Figure 6: Adsorption isotherm of sodium dodecylsulfate (SDS) on alumina. Surfactant aggregation numbers determined at various adsorption densities are shown along the isotherm

The concept of surfactant aggregation at the solid-liquid interface has been employed by several authors, to account for the sharp changes in interfacial properties such as the amount adsorbed, hydrophobicity and zeta potential observed above a critical surfactant concentration. This process was termed hemimicellization by analogy to micellization. Lateral aggregation, for example, among adsorbed dodecylsulfate species on alumina has been shown to result in drastic increase in the adsorption density, settling rate as well as the electrophoretic mobility[20] (figure 7). The hemimicellar aggregates that form in region II were viewed as two-dimensional monolayered structures with the tails extended towards the solution side. The molecular structure of the adsorbed layer itself has been the subject of often controversial speculation. Spectroscopic probing using fluorescence, ESR and Raman methods were designed to yield information on the microstructure of the adsorbed layer.

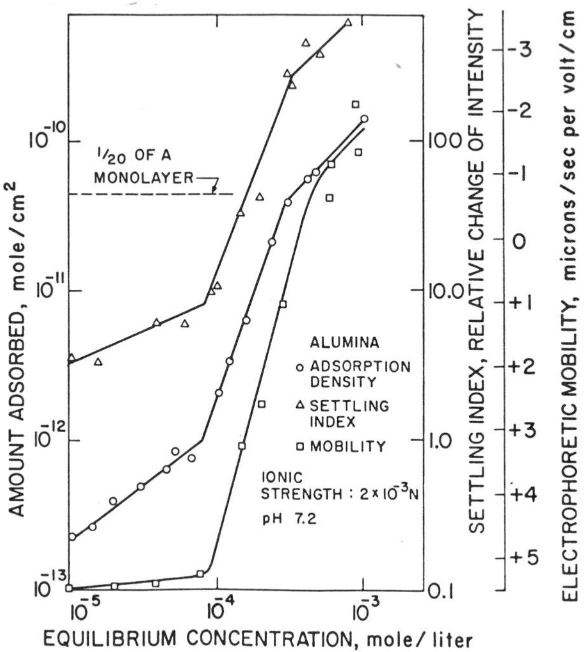

Figure 7: **Changes in interfacial properties of alumina as a result of sodium dodecylsulfate adsorption**

Fluorescence Studies of Surfactant Aggregation at the Alumina-water Interface[6]

Fluorescence emission of pyrene in micellar solutions can be used to obtain information on the aggregation number of surfactants. For this the decay kinetics of the monomer and excimer emissions of pyrene at different concentration levels are determined and a kinetic analysis based on the relations connecting the decay rates of the monomer and excimer is carried

 Characterization Techniques for the Solid-Solution Interface

out from the decay profiles of pyrene in the sodium dodecylsulfate (SDS) layers adsorbed on alumina at different adsorption densities. It was confirmed first that the micromolar concentrations of pyrene used in our experiments did not affect the adsorption of SDS on alumina. The aggregation numbers thus obtained are marked on the adsorption isotherm in figure 6. The aggregates in region II appear to be relatively of uniform size, but in region III there is a marked increase in the aggregate size. These results give an idea of the evolution and structure of the adsorbed layer. Region II and above seem to be characterized by surfactant aggregates of limited size. Since the surface is not totally covered, there are enough positive sites available for SDS and adsorption occurs mainly by increasing the number of aggregates rather than the size of each aggregate. The transition from region II to III corresponds to the iso-electric point of alumina and adsorption in this region occurs through the growth of the existing aggregates. A schematic representation of the evolution of the adsorbed layer is provided in figure 8.

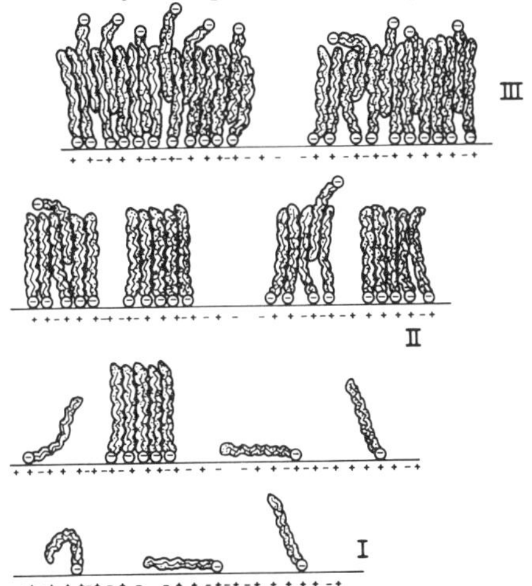

Figure 8: **Schematic representation of the growth of sodium dodecylsulfate aggregates at the alumina-water interface**

Electron Spin Resonance Studies of Surfactant Adsorption and Aggregation

Stable free radical nitroxide spin probes were chosen for our ESR studies. They were the three isomeric 5-, 12- and 16- doxyl stearic acids (5-D, 12-D, 16-D). These spin labels were coadsorbed individually on alumina along with the main adsorbate, sodium dodecylsulfate (SDS), and the regions of the adsorption isotherm were studied. The responses of 16-D along the adsorption isotherm are shown in figure 9. The broad spectrum obtained at low adsorption densities is characteristic of the spin-spin relaxation occurring when two nitroxides interact with each other (spectrum A). As the adsorption of the dodecylsulfate among the probe increases and its aggregates form, probe-probe interactions decrease and a sharper anisotropic spectrum is obtained (spectrum C) which indicates greater mobility of the probe. The spectra of 16-D was compared with the spectra of the probe in ethanol-glycerol mixtures of known viscosities and a microviscosity of 120-165 cP was estimated for the SDS aggregates.

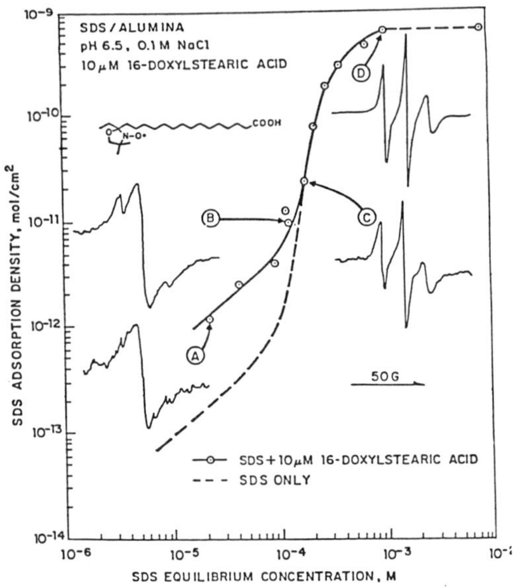

Figure 9: **ESR spectra of 16-doxyl stearic acid in sodium dodecylsulfate - alumina slurries along various regions of the adsorption isotherm**

Figure 10 compares the spectra of the 5-, 12- and 16- doxyl stearic acid obtained in SDS hemimicelles with those in ethanol/glycerol mixtures and it is evident that the nitroxide closer to the surface is more immobile than the one farthest from it. This implies that the chain segments near the surface are tightly packed while those near the end of the chain are considerably more disordered. Thus ESR spectroscopy with nitroxide spin probes is demonstrated here to be an adept method for probing the microstructure of adsorbed layers.

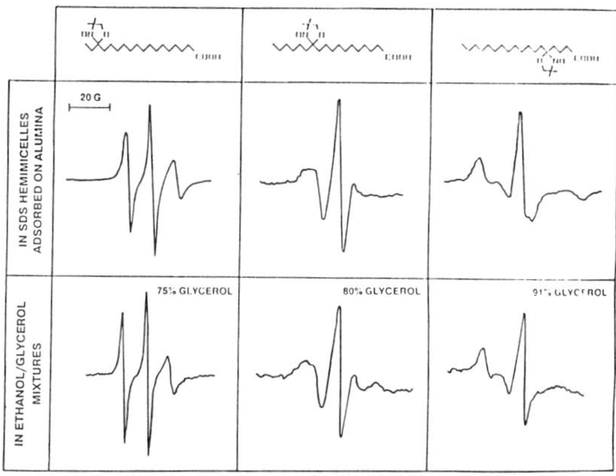

Figure 10: **ESR response of the three probes (5, 12 & 16-doxyl stearic acid) in sodium dodecylsulfate hemimicelles adsorbed on alumina and in ethanol/glycerol mixtures. From left to right, the probe describes an environment increasingly closer to the alumina surface.**

Time Resolved Resonance Raman Studies

The Raman spectra of the probe in sodium dodecylsulfate aggregates at the alumina-water interface corresponding to different regions along the adsorption isotherm are shown in figure 11. The spectra obtained in regions I and II remain similar to that obtained in aqueous spectrum. The spectra in regions III and IV show variation in intensity and frequency, indicating a definite trend in the magnitude of the changes. The shift in frequencies and variation in intensities of some of the Raman lines for various regions of the isotherm almost trace the shape of the isotherm when plotted as a function of residual concentration. This suggests that the sensitivity of the Raman probe to changes in the hemimicellar environment can be exploited to obtain information on the structure of adsorbed layers.

ESR IN NON-AQUEOUS MEDIA

The traditional medium for the processing of ceramic powders has been water but in some recent developments, organic media have been preferred to eliminate some of the defects arising in aqueous media[21]. For instance, ceramics of current interest such as nitrides and carbides, which tend to develop oxide surfaces when processed in water are better processed in nonaqueous media. Nonaqueous liquids are used also when drying is a particular problem such as in casting electronic substrates. They are also favored when the dispersion of the powder material requires use of processing additives of limited solubility in water.

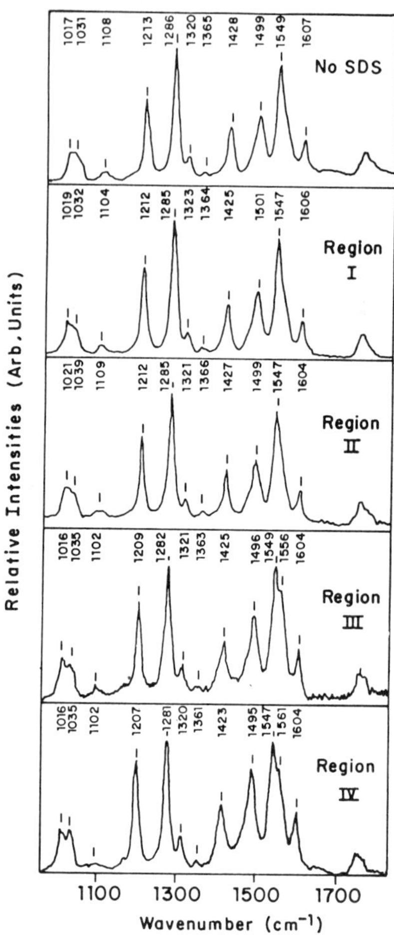

Figure 11:   Resonance Raman spectra of Ru(bpy)$_3^{2+}$ on alumina slurry, and along various regions of the sodium dodecylsulfate/alumina adsorption isotherm

Steric stabilization has been considered recently as a major mechanism in non-aqueous dispersions. The mechanisms governing the steric stabilization are rather well understood, especially when using macromolecular additives.[22,23] However, the choice of stabilizer for a particular system has been empirical. Recently we have used spectroscopic techniques to investigate the conformation of molecules at the solid-liquid interface and its role in determining the efficiency of interfacial phenomena such as

    Characterization Techniques for the Solid-Solution Interface

aggregation/dispersion of colloidal suspensions in non-aqueous media.[24,25] In non-aqueous media, line broadening due to spin-spin relaxation, that was observed in aqueous media, was not detected in experiments performed with 5, 12 and 16-D stearic acids adsorbed on alumina in cyclohexane. This is direct evidence for the fact that in non-polar media, surfactant molecules do not interact laterally as in the case of aqueous solutions where surfactant hemimicelles are formed at the alumina-water interface.

On the other hand, when the probe is coadsorbed with Aerosol OT, the stearic acid molecules show significant changes in their ESR line shape as indicated in figure 12. At low surface coverage, interactions between the probe and the alumina surface leads to a "frozen" spectrum due to slow rotation of the molecules. When Aerosol OT is added, its adsorption induces sufficient surface pressure to push the nitroxide away from the surface in the case of 12 and 16-D stearic acid, allowing the molecules to dangle out in a stretched conformation.

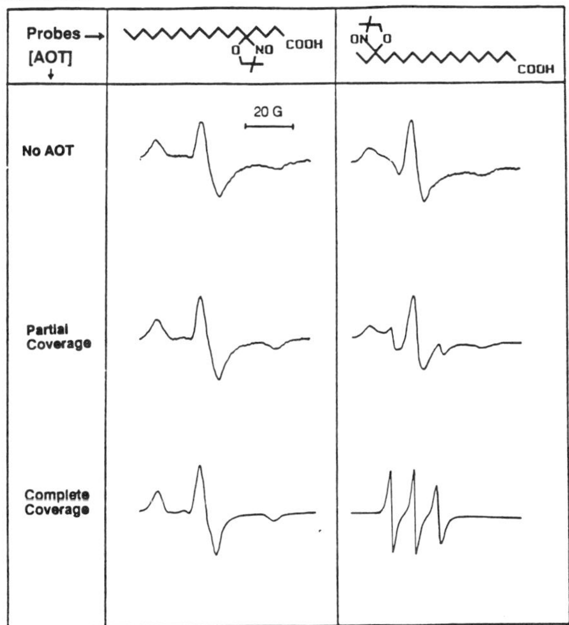

Figure 12:   **ESR spectra obtained with 5- and 16-doxyl stearic acids adsorbed at the alumina/cyclohexane interface in the presence of various amounts of Aerosol OT.**

In figure 13 changes in the ESR response of 12-doxyl stearic acid are shown as a function of Aerosol OT adsorption density. It is to be noted that using changes in the ESR signal intensity at a given position on the ESR spectrum, it is possible to follow the kinetics of surfactant adsorption nearly instantaneously and continuously, an experiment not easily performed by any technique.

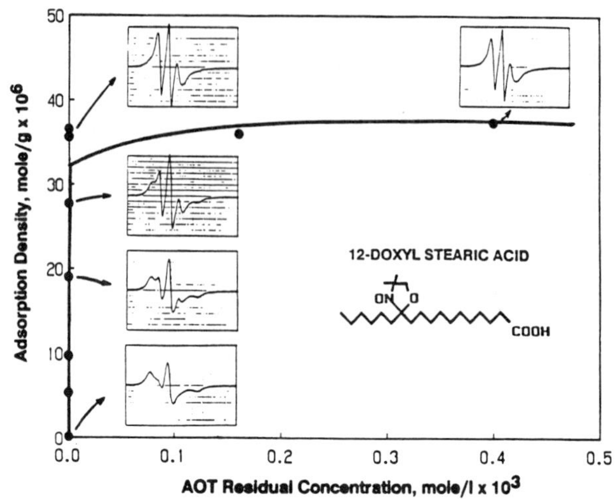

Figure 13:   **Adsorption isotherm of Aerosol OT on alumina in cyclohexane and corresponding ESR spectra obtained using 12-doxyl stearic acid as a probe**

As can be seen from figure 14, among other changes in the spectrum, a third peak appears and progressively increases in intensity as the surfactant adsorption is increased. This peak corresponds to the highly mobile population of probe molecules dangling in solution by the coadsorption of surfactant around them. It was arbitrarily selected as an indicator of surfactant adsorption since it changes significantly during surfactant adsorption. The overall adsorption phenomenon is relatively fast and 80% of the total adsorption was complete within the first 10 minutes after surfactant addition.

The above results illustrate the wide range of information ESR can provide a ceramic chemist and their relevance to fundamental investigations of colloidal dispersions in non-aqueous media: it offers an unique means to study surfactant adsorption at interfaces and allows quantitative measurement of the packing density of adsorbed layers, a critical parameter for steric stabilization. For most of the above mentioned applications, ESR spectroscopy does have some inherent disadvantages that we tried to minimize. The probes used reflect the properties of the environment in which it resides, but it also perturbs to a certain extent as can be seen from figure 9. Another limitation of this technique is that it can be used only for systems that have extremely low concentrations of free radicals and other paramagnetic impurities. Presence of transition metal ions such as Fe, Cu and Ti either as soluble species or in the mineral itself limits the use of this technique. Finally, extreme caution must be exercised while obtaining quantitative information from the spectra because the ESR signal intensity does not depend only on the probe concentration in the system, but it also varies according to the location of the sample in the spectrometer cavity, the presence of impurities in the cavity and sample.

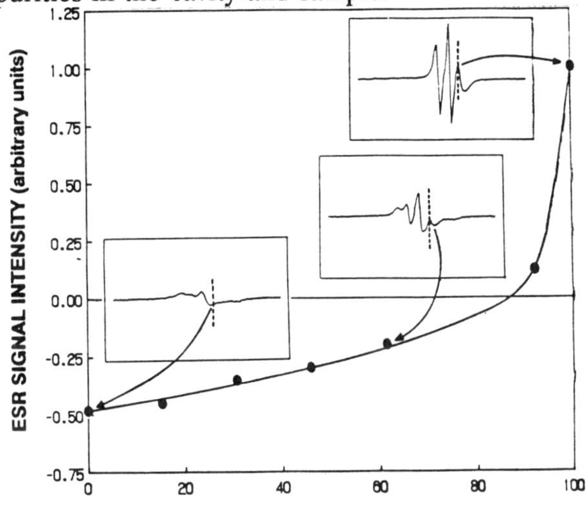

Figure 14: **Intensity of the ESR peak (indicated by the dotted line on the spectra in insets) as a function of surface coverage from the adsorption isotherm data**

Infra-red Spectroscopy at the Solid-Liquid Interface

In separation of francolite from dolomite by flotation using potassium oleate as the collector, it was observed that selective separation expected from single mineral tests could not be achieved when using a mixture of the minerals[26]. This was attributed to the non-selective precipitation of calcium and magnesium oleate species on the mineral surfaces, thereby altering their interfacial properties. The surface chemical characteristics of oleate precipitates were examined using FTIR/ATR spectroscopy. At low oleate concentrations, there was individual oleate molecule adsorption on the mineral surface without any precipitation; as the oleate concentration was further increased, surface precipitation of Ca and Mg oleates occurred inducing hydrophobicity on the mineral surfaces; at high oleate concentrations, interaction with the dissolved mineral species caused bulk precipitation, leading to a loss of flotation selectivity and resulted in high reagent consumption.

ROLE OF POLYMER CONFORMATION IN COLLOIDAL STABILITY

Dispersion or flocculation of suspended particles is determined by macromolecular adsorption both in terms of the amount adsorbed and the configuration of the adsorbed species. Even though it has been recognized that configuration of polymers at the interfaces can lead to either flocculation or dispersion, it has never been established as to what type of conformation (stretched versus coiled or flat versus dangled) is optimal for the above or as to how one can manipulate a system to achieve the optimum configuration. This lack of knowledge has been essentially due to the non-existence of reliable in-situ techniques with which one can determine conformation and orientation of species adsorbed on solids in liquids. Techniques such as gel permeation chromatography, light scattering and viscosity can give information on the average conformation of the polymer in the bulk but are incapable of doing so at the solid-liquid interface. Recently we have developed a multi-pronged approach involving simultaneous measurements of dispersion/flocculation responses and configuration of adsorbed polymer species using luminescence spectroscopy of pyrene-labeled polyacrylic acid[27].

Dispersion/flocculation of alumina suspensions by polyacrylic acid was studied and the effect of polymer conformation on the suspension stability was determined. The conformation of the polymer at the solid/liquid interface was monitored as a function of pH. At low pH (~4) the adsorbed polymer was found to be coiled while at high pH (~8), it was stretched. The conformational state of the polymer in solution was same as that adsorbed on alumina.

In our efforts to manipulate the polymer conformation at the solid/liquid interface for dispersion control, it was discovered that if the polymer is adsorbed first at low pH, and then the pH raised, excellent solid-liquid separation is obtained (Figure 15a). On the other hand, if the polymer was adsorbed on alumina at pH 10 and then decreased to 4, the flocculation responses were not as significant as those obtained in figure 15 but were as bad as those found under fixed pH conditions (figure 15b). Comparing this to the system under fixed pH conditions (Figure 16) it can be concluded that a change in adsorbed polymer conformation from coiled to slightly extended will result in best settling results. A schematic representation of the conformational changes of the polymer is shown in figure 17 a-f. Starting from pH 4, the coiled polymer in solution (uncharged polymer) adsorbs and remains in the same coiled conformation on the positively charged alumina surface (17a). When the pH is raised to the intermediate neutral range (~5-7), ionization of PAA generates some negative charges on the polymer, and causes the polymer chain to expand.

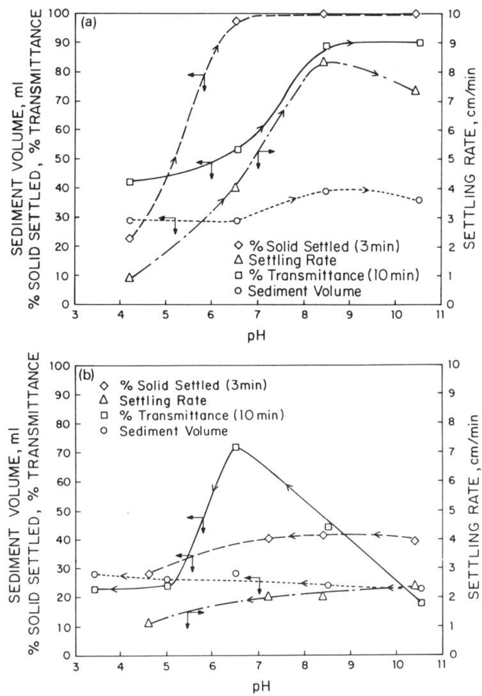

Figure 15: Flocculation properties of alumina with 20 ppm PAA as a function of final pH under changing pH conditions with the initial pH of 4 (a) and 10 (b)

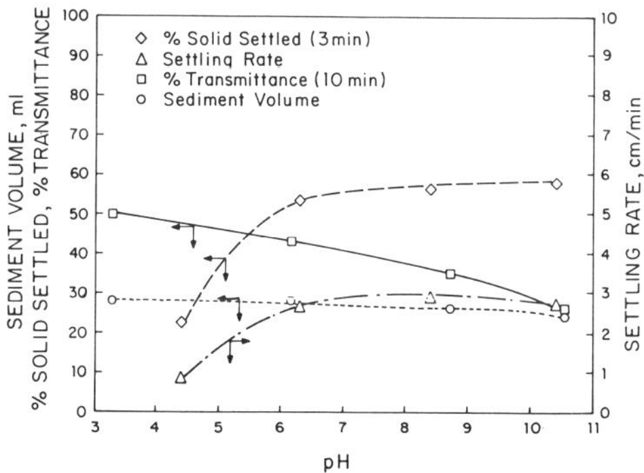

Figure 16: Flocculation of aqueous alumina suspensions with 20 ppm PAA

The alumina is still positively charged and the polymer will adsorb due to electrostatic attraction (17b). Raising the pH above the point of zero charge of alumina (pH 8.5), the solid particles become negatively charged similar to the polymer and under these conditions, electrostatic repulsion would cause the polymer to be displaced away from the particles and "dangle" from the particle into the aqueous phase (17c). In contrast, when the suspension is first treated with PAA at pH 10, the polymer is probably weakly adsorbed on the surface through hydrogen bonding (17d). Reducing the pH to below the pzc of alumina, the charge on the particles becomes positive and increases attraction with the polymer. This may result in the PAA adsorbing in a flat conformation to maximize the contact between the polymer and mineral (17e). Further reduction in the pH to the region where PAA is neutral does not alter the conformation of the adsorbed polymer (17f). The conformation in figure 17b and c is most suited to flocculation hence the observed performance when the pH of the adsorbed polymer is raised from 4 to 10. On the other hand, the conformation in figure 17 e and f is least suited for flocculation hence there is no increase in flocculation when the pH is lowered from 10 to 4.

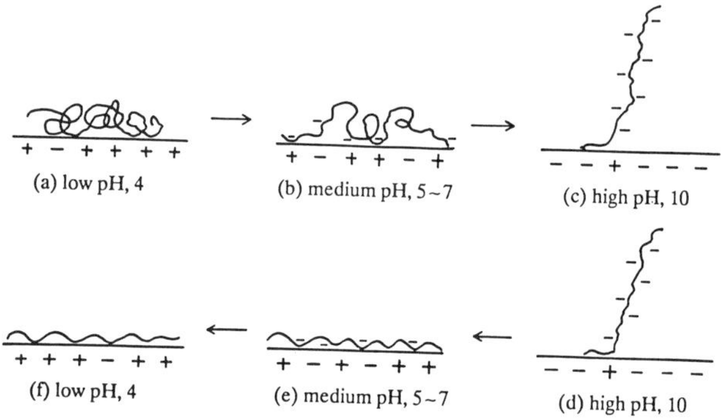

Figure 17:   **Schematic representation of variation in polymer conformation at the alumina-solution interface under changing pH conditions**

POTENTIAL OF NUCLEAR MAGNETIC RESONANCE (NMR) SPECTROSCOPY

Fluorescence and ESR spectroscopy require the use of external probes: such probing has demonstrated its usefulness in providing information unobtainable by other means but, no matter what precautions are taken, there will always be a question on the extent to which the probe perturbs its environment. This is why non invasive techniques are preferred when available. NMR is a spectroscopic technique that has this non-invasive quality. Recent advances in NMR instrumentation has made it possible to study various reagents, at very dilute concentrations and even at the solid-liquid interface. In some exploratory work, we have established that this can be used to study adsorbed layers at the solid-liquid interface, both in aqueous and non-aqueous media. Even though there was reagent in the supernatant proper set-up of the parameters can eliminate interference from signals in bulk solution so that only the adsorbed layer is probed. It will not be long before this technique will be applied to more in-depth studies at the solid-solution interface and take research in interfacial technology to an entirely new level.

SUMMARY

Structural characterization studies are of immense help for understanding the nature of the adsorbed layers and thus to provide a better opportunity for manipulating and controlling interfacial properties for such diverse applications as ceramic processing, magnetic tape manufacturing, liquid inks, paints, cosmetics and oil recovery from tar sand. This article summarizes some of the recent results obtained in this laboratory on the *in-situ* characterization of adsorbed layers at the ceramic solid - solution interface in aqueous as well as nonaqueous solutions using fluorescence spectroscopy, electron spin resonance spectroscopy and resonance raman spectroscopy. The approach used here can be used for better prediction of suspension behaviors and offers new techniques to assess steric stabilization. Control of colloidal stability by in-situ manipulation of adsorbed polymer conformation has been clearly demonstrated.

ACKNOWLEDGEMENT

Financial support from the National Science Foundation, Department of Energy, NALCO Chemical Company, Engelhard Corporation, BP America Inc, Unilever Research U.S. Inc. and ARCO Oil and Gas Company is acknowledged.

REFERENCES

1.   J.Th.G. Overbeek, "Recent Developments in the Understanding of Colloid Stability", *J. Colloid Interface Sci.*, **58(2)**, 408-422, 1977.

2.   I.A. Aksay, F.F. Lange and B.I. Davies, "Uniformity of Al_2O_3 - ZrO_2 Composites by Colloidal Filtration", *J. Amer. Ceram. Soc.,* **66**, 1983.

3.   A. Bleier, "Stability of Ceramic Suspensions", 391-403, in <u>Ultrastructure Processing of Ceramics, Glasses and Composites</u>, L.L. Hench and D.R. Ulrich (eds.), John Wiley and Sons, New York, 1984.

4.   P. Somasundaran and H.S. Hanna, "Physico-Chemical Aspects of Adsorption at Solid/Liquid Interfaces, Part 1 - Basic Principles", 205-252, in <u>Improved Oil Recovery by Surfactant and Polymer Flooding</u>, D.O. Shah and R.S. Schechter (eds.), Academic Press, New York, 1977.

5.   J.P. Lakowicz, <u>Principles of Fluorescence Spectroscopy</u>, Plenum Press, New York, 1983.

6.   P. Chandar, P. Somasundaran and N.J. Turro, "Fluorescence Probing Studies on the Structure of Adsorbed Layer of Dodecylsulfate at the Alumina-water Interface", *J. Colloid Interface Sci.,* **117**, 31, 1987.

7.   P. Chandar and P. Somasundaran, "Excimer Fluorescence Determination of Solid-Liquid Interfacial Pyrene-labeled Polyacrylic acid", *Langmuir,* **3**, pp 298-300, 1987.

8.   J.K. Thomas, "The Chemistry of Excitation at Interfaces", ACS Monograph, American Chemical Society, Washington, D.C., 1984.

9.   R. Zana, "Luminescence Probing Methods", 241-294, in <u>Surfactant Solutions - New Methods of Investigation</u>, R. Zana (ed), Surfactant Science Series, **22**, Marcel Dekker, New York, 1987.

10.   H.T. Oyama, D.J. Hemker and C.W. Frank, "Effect of the degree of ionization of Poly (methacrylic acid) on Complex formed with Pyrene end-labeled Poly (ethylene glycol)", *Macromolecules,* **22 (3)**, 1255-1260, 1989.

11.   L.J. Berliner, <u>Spin Labelling I: Theory and Applications</u>, Academic Press, New York, 1979.

12.   C.A. Malbrel, "Effect of Water on the Dispersion of Colloidal Alumina in Cyclohexane Solutions of Aerosol OT", Doctor of Engineering Science Thesis, Columbia University, New York, 1991.

13. P. Somasundaran and J.T. Kunjappu, "In-situ Investigation of Adsorbed Surfactants and Polymers on Solids in Solution", *Colloids and Surfaces*, **37**, 245-268, 1989.

14. P.R. Griffiths and J.A. de Haseth, <u>Fourier Transform Infrared Spectrometry</u>, Chemical Analyses Series **vol. 89**, John Wiley & sons, New York, 1989.

15. M. van der Waarden, "Stabilization of Carbon Black Dispersions in Hydrocarbons", *J. Colloid Interface Sci.*, **5**, 317-325, 1950.

16. P. Somasundaran, T.W. Healy and D.W. Fuerstenau, "Surfactant Adsorption at the Solid-Liquid Interface - Dependence of Mechanism on Chain Length", *J. Physical Chem.*, **68**, 3562, 1966.

17. P. Somasundaran and D.W. Fuerstenau, "Mechanism of Alkyl Sulfonate Adsorption at the Alumina-water Interface", *J. Physical Chem.*, **70**, 90, 1966.

18. P. Somasundaran, P. Chandar, N.J. Turro and K.C. Waterman, "Investigations into the Structure of Adsorbed Layer of Dodecylsulfate at the Alumina/water Interface", 775, in <u>Proceedings of XVIth International Mineral Processing Congress</u>, K.S.E. Forssberg (ed), Stockholm, Sweden, 1988.

19. P. Chandar, "Fluorescence and ESR Spectroscopic Studies of the Structure of Dodecylsulfate Adsorbed Layer at the Oxide-water Interface", Doctor of Engineering Science Thesis, Columbia University, New York, 1986.

20. P. Somasundaran, T.W. Healy and D.W. Fuerstenau, "The Aggregation of Colloidal Alumina Dispersions by Adsorbed Surfactant Ions", *J. Colloid Interface Sci.*, **22**, 599, 1966.

21. J.S. Reed, <u>Introduction to the Principles of Ceramics Processing</u>, John Wiley and Sons, New York, 1988.

22. T. Sato and R. Ruch, <u>Stabilization of Colloidal Dispersions by Polymer Adsorption</u>, Marcel Dekker, New York, 1980.

23. D.H. Napper, <u>Polymeric Stabilization of Colloidal Dispersions</u>, Academic Press, England, 1983.

24.   C.A. Malbrel, P. Somasundaran and N.J. Turro, "Adsorption of Nitroxide Spin Probes at the Alumina/Cyclohexane Interface in the Presence of Aerosol OT", *Langmuir*, **5**, 490, 1989.

25.   C.A. Malbrel and P. Somasundaran, "In-situ Kinetics Measurements of Surfactant Adsorption on Colloidal Alumina using ESR Spectroscopy", *J. Colloid Interface Sci.*, **137** (2), 600-603, 1990.

26.   P. Somasundaran, L. Xiao and D. Wang, "Solution Chemistry of Sparingly Soluble Minerals", *Minerals and Metallurgical Processing*, **8** (3), 115-121, 1991.

27.   K.F. Tjipangandjara, Y.B. Huang and P. Somasundaran, "Correlation of Alumina Flocculation with Adsorbed Polyacrylic acid Conformation", *Colloids and Surfaces*, **44**, 229-236, 1990.

Characterization of the State of Dispersion of Particle Suspensions

CHARACTERIZATION TECHNIQUES FOR AGGLOMERATE STRUCTURES

Brij M. Moudgil and Sanjay Behl
Mineral Resources Research Center
University of Florida, Gainesville, Fl

ABSTRACT

Presence of hard agglomerates in ceramic slurries is known to have an adverse impact on the product properties. Until recently presence of soft agglomerates (flocs) was also undesired. Lately, weak flocculation of ceramic slips is reported to be effective in preventing particle segregation and result in faster casting rate and higher green densities under appropriate conditions. The properties of the soft agglomerates (flocs) such as size, distribution, shear strength and deformability determine the effectiveness of flocculation in the processing step. Considering significant differences in size, density and shear strength of hard and soft agglomerates, not all of the techniques developed for characterizing the former can be used to characterize soft agglomerates. In this paper, techniques developed for determining size, distribution, density and shear strength of soft agglomerates (flocs) are described.

INTRODUCTION

The aggregation/dispersion of particles is an important parameter in colloidal ceramic processing. Aggregation involves the agglomeration of fine particles by reducing the zeta potential to zero, increasing electrolytes concentration or additions of organic macromolecule/polymers to form "flocs" or soft agglomerates. Presence of flocs traditionally was considered to have a deleterious effect on the sintered ceramic properties however, more recently they have shown to be advantageous in the processing step (1), particularly in enhancing the casting rate and for minimizing segregation of two components as different densities. The major applications of these structures is in enhancing the casting rates. Presence of agglomerates (flocs), by properly adjusting the conditions for formulation of a ceramic slip, can yield higher green densities [1]. The floc properties such as size, size distribution, shear strength and deformability, play an important role in determining the effectiveness of agglomeration in the processing step. For example, in filtration the "flocs" need to be highly porous and large in size, on the other hand if the process requires high shear applications (like injection molding) then the flocs need to be able to handle the shear without breaking or disrupting. Proper characterization of the flocs, therefore, is essential for understanding the effect of these structures or particle networks in a ceramic slip.

Two type of floc structures, open or compact as illustrated in Figure 1, can be encountered. Generally, Compact flocs are generated by coagulation and open flocs with polymer induced flocculation. Sutherland [2], based on simulation results, proposed two types of mechanisms: 1) aggregate growth by addition of single particles to a growing seed cluster, like addition polymerization and 2) aggregation by the collision of clusters as in Smoluchowski theory of rapid coagulation or LaMer's model for flocculation. It was shown by Sutherland [2] that in the first case the flocs were compact while in the latter case they were open structures. Clearly classification of these clusters would pose a number of challenges primarily because as the floc size increase there is more open voidages, thereby decreasing the density and giving flocs a fractal characteristics. In this paper experimental techniques to characterize flocs or soft agglomerates with respect to size, size distribution, density, shape and shear strength are discussed. Most of the techniques are similar for characterizing hard and soft agglomerates. However, the shear strength of the hard agglomerates is considerably higher than of flocs, the size of the flocs on the other hand is relatively larger than hard agglomerates.

SIZE, DISTRIBUTION AND DENSITY

The floc size and density are studied together because of the fractal nature of the flocs. Also most of the size measurement techniques consider both the size and the density at the same time. The measurement of floc size is critical to nearly all the other floc characterization aspects. The size may be measured using any of the analytical techniques outlined in the literature for similar measurements for hard agglomerates. However the techniques using Stokes equation for giving the equivalent size, may be erroneous as floc density is a function of size and needs to be measured directly.

A number of methods have been proposed for the measurement of size and density either separately or together. These can be summarized as (1) hindered settling, which gives an average floc density from the volume of settled sediment and an average floc size from the rate of fall of the mud line, (2) Koglin's [3] method of using simultaneous turbidity and sedimentation balance measurements on the suspension as a whole (3) Klimpel's [4] method of measuring from the strobe flash photography the velocity of fall and the size of individual floc (4) Warren's instruments [5] for in-situ measurement of both floc density and size at the same time. Floc size distribution may also be measured by using a Cahn sedimentation balance [6] once the relation between the density and size is established. A comparison of some these methods with simulation results is given in Figure 2 [5]. It is interesting to note that depending on the method used, the actual values

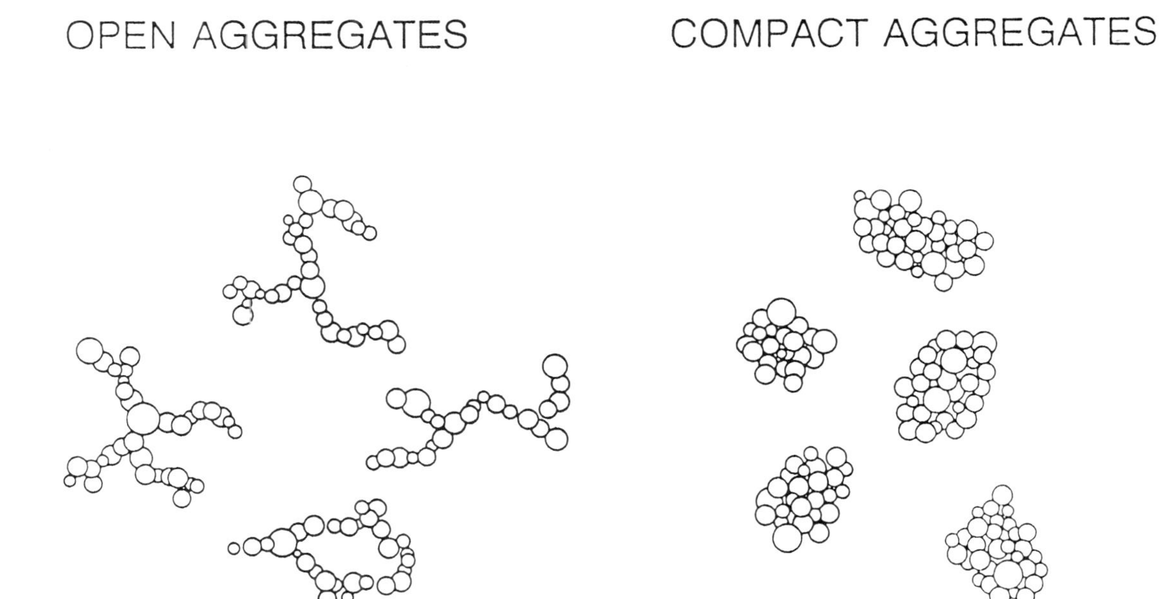

Figure 1. The two extremes of floc structure: "open" aggregates and "compact" aggregates.

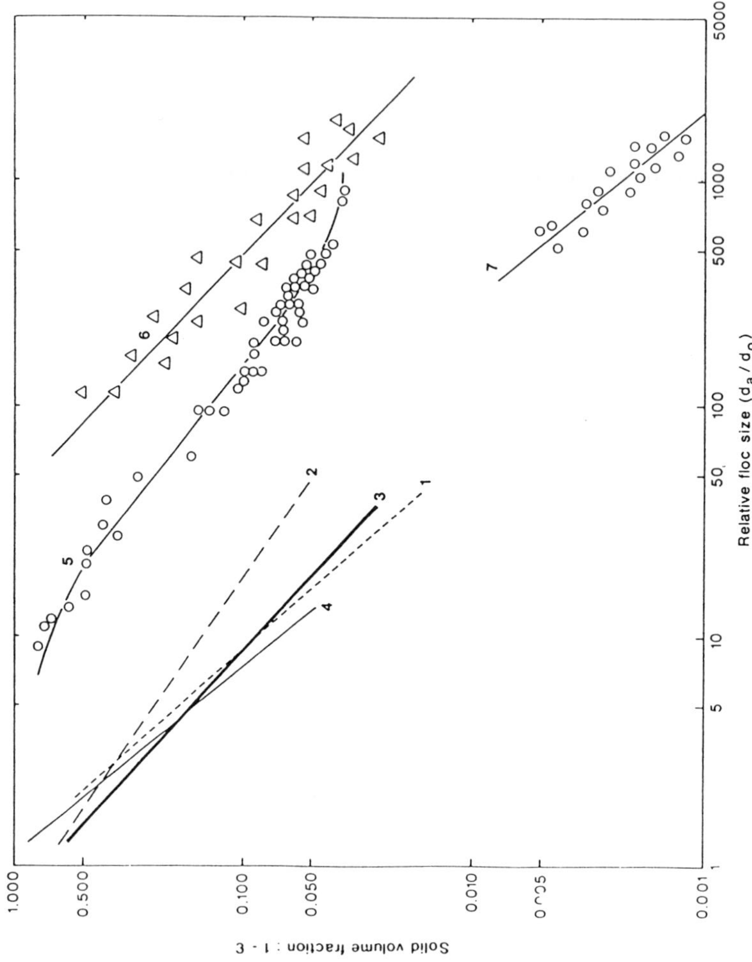

Figure 2. Solid volume fraction as a function of relative floc size from theory: (1) Goodarz-Nia, [8], (2) Vold [9], (3) direct measurements, Farrow and Warren [5], (4) Koglin [3], (5) Klimpel et al [4], (6) Dirican [6], and (7) Tambo and Watanabe [7].

obtained are different. However the slopes for density and size plot are fairly constant, implying that the floc formation process can be considered to be a fractal growth process. Some of the flocs formed in these studies were by coagulation and others by polymer induced flocculation.

The density of a floc is usually evaluated by measuring its buoyant density while settling in a quiescent column. An equation of the general form was used by Matsumo and Mori [10] calculate the buoyant density from settling rate and diameter of the floc. The constants α,β,γ depend on the shape, porosity and the flow conditions around the floc. It is known that shape and porosity of flocs vary widely, consequently it is difficult to establish meaningful relationship between the constants.

$$U = \frac{K_f g \, \rho_a^{\alpha} D_f^{\beta}}{\mu^{\gamma}}$$

A number of attempts to evaluate floc density have been reported in the literature (Matsumo and Mori [10], Tambo and Watanabe [7], Dirican [6], Klimpel [4], and Hogg [11]). However different equations were used to calculate the floc densities and none of these studies accounted for shape of the flocs in an appropriate manner (see Table 1). In most of the cases the equations used to represent the settling rate of the flocs apparently were selected on the basis of least statistical error obtained by evaluating the power law relationship between diameter of the floc and its density.

Study	Settling Rate U	Drag Coefficient C_D
Matsumo and Mori [10]	$K_f \rho_a^{0.85} D^{1.54} / \mu^{0.7}$	$K / N_{Re}^{0.82}$
Tambo and Watanabe [7]	$K_f \rho_a^{0.5} D^{0.5} / \rho_w^{0.5} C_D^{0.5}$	K / N_{Re}
Dirican [6], Klimpel and Hogg [13], and Hogg, et al [14]	$K_f \rho_a^{0.5} D^{0.5} / \rho_w^{0.5} C_D^{0.5}$	$K[1+K' / (N_{Re})^{0.5}]^2$
Moudgil and Vasudevan [12]	$K_f \rho_a^{0.5} D^{0.5} / \rho_w^{0.5} C_D^{0.5}$	K / N_{Re}

Table 1. Settling rate equations used in various studies

In the present investigation, floc size was measured using an optical microscope. Such techniques can be automated by using an image analyzer. It should be noted that the floc size measurement by optical microscopy yields the projected view of the three dimensional aggregate.

Procedures for measurement of floc size and density

The experimental technique for measurement of floc size and distribution by microscopic method involves transfer of the flocculated mass from a mixing vessel into a large tray to prevent reagglomeration, followed by scoping a few agglomerates on a glass slide. The water surrounding the flocs is carefully blotted out using an absorbent tissue paper. No change in the floc size or shape was observed before and after blotting out the water [12]. The slide is place under the microscope provided with a grid and the projected area is measured.

The density may be determined using either direct measurement or sedimentation techniques. Direct measurements involve removal of water surrounding the floc by the blotting paper, followed by determination of its mass before and after drying for 4 hours at 60°C. The volume fraction of the solids and water in the flocs and thus the floc density can be calculated using simple mass balance. It should be noted that direct measurements would be reliable only for larger flocs, for smaller flocs sedimentation method is used to obtain more reliable data.

In the sedimentation technique, the projected area of the flocs is first measured using a microscope. The terminal velocity is then determined by settling in a 70 cm column of water (4.6 cm ID). The Stoke's relationship (outlined in the next section) is used to calculate the floc density.

Illustration

In order to establish the reliability of the above procedure a comparison of floc density directly obtained for a kaolinite suspension flocculated with non-ionic polyacrylamide in the present study was made with the values reported in a previous study [14] by sedimentation method (See Table 2). Reasonable agreement between the values obtained indicates the reliability of the two density measurement techniques.

SHAPE

The settling rate for larger flocs is typically used to measure the size of the flocs. A modified Stoke's equation which relates the terminal settling velocity (V) with buoyant density and size of a given floc (d) is generally used.

$$V_t = \frac{4g\, d_f\, (\rho_f - \rho_w)}{3\, C_D\, \rho_w}$$

Where d_f is floc diameter and ρ_f and ρ_w are the density of floc and water respectively.

Floc Diameter, μm	Floc Density, g/cm^3	
	Present Study	Hogg, et al [14]
945	1.035	1.024
1000	1.033	1.022
1130	1.023	1.022

Source:	hogg, et al 1987
System:	3 wt% Kaolinite Suspension
Polymer:	Present Study: 15 million mol.st. non-ionic polyacrylamide
	Previous Study: 13 million mol.st. non-ionic polyacrylamide
Polymer Dosage:	Present Study: 0.27 kg/t solids
	Previous Study: 0.07 kg/t solids
Time of Flocculation:	60 sec (in both studies)
Agitation Intensity:	1000 rpm (in both studies)

Table 2. A comparison of floc density values obtained in the present study with a previous investigation.

The effect of liquid flow field around the floc is accounted for by the drag coefficient term Cd which is related to the floc Reynolds number, its shape and porosity. In the past studies, calculation of the drag coefficient has generally assumed that the flocs as rigid spheres, thus ignoring the effect of porosity and shape. Matsumo and Suganuma [15], in examining the effect of porosity on settling rate of a model floc made of steel wool, showed that the flocs with solids volume fraction higher than 0.05 indeed behave as rigid bodies. This observation was confirmed recently by Moudgil and Vasudevan [12] by comparing the settling rates of frozen and unfrozen kaolinite flocs of similar size. Most of the past researchers also assume flocs to be of spherical shape. However Moudgil and Vasudevan showed that this may no always be true and used the following shape factor as described by Brown.

$$\psi = \frac{\text{Surface Area of Sphere Equivalent in Volume to Floc}}{\text{Surface Area of Floc}}$$

The significance of the shape factor in density determination by Stoke's equation is illustrated below.

Illustration

The shape of dolomite and kaolinite flocs was found to be ellipsoidal, and the shape factor of these flocs by above relationship was measured to be 0.8. The floc densities calculated using different values of shape factor, are shown in Figures 3 and 4. The densities, calculated assuming flocs to be spherical are much smaller than those obtained by direct measurements indicating that in using settling rate technique for evaluation of the floc density an improper assumption with respect to shape of flocs can lead to highly erroneous results.

FLOCS AS FRACTALS

It is well known that aggregate of colloidal particles exhibit fractal characteristics. A cluster - cluster aggregation model was developed by Meakin to describe the aggregation process [13]. Two primary mechanisms namely, diffusion limited and reaction controlled were suggested. The fractal dimension of the floc may or may not represents the mechanism for cluster formation. However the fractal dimension number can be used to classify flocs for compactness.

Dolomite and kaolinite suspensions were flocculated using Polyethylene Oxide (PEO) as a flocculant. It can be seen from Figures 5 and 6 that dolomite flocs yield a fractal dimension of 1.68 while kaolinite has a value of 2.04 indicating that the dolomite flocs are less compact than kaolinite flocs.

FLOC SHEAR STRENGTH

Most of the past studies to determine the aggregate strength were conducted under viscous flow conditions. However it is more appropriate to determine the floc strength under the turbulent conditions that would be experienced during most of the industrial processing steps. Glasgow and Hsu evaluated floc strength by using a turbulent air jet to break the flocs and measuring the resulting floc size [17]. Since it is difficult to scale-up the data in terms of hydrodynamic conditions it is more useful to measure floc strength by breaking the flocs under the turbulent flow in the standard mixing tank. The binding strength of a floc was therefore determined by breaking the flocs under a known shear field and measuring the resulting size of the flocs formed. The binding strength of the floc formed is calculated from the applied shear force applied using the following equation.

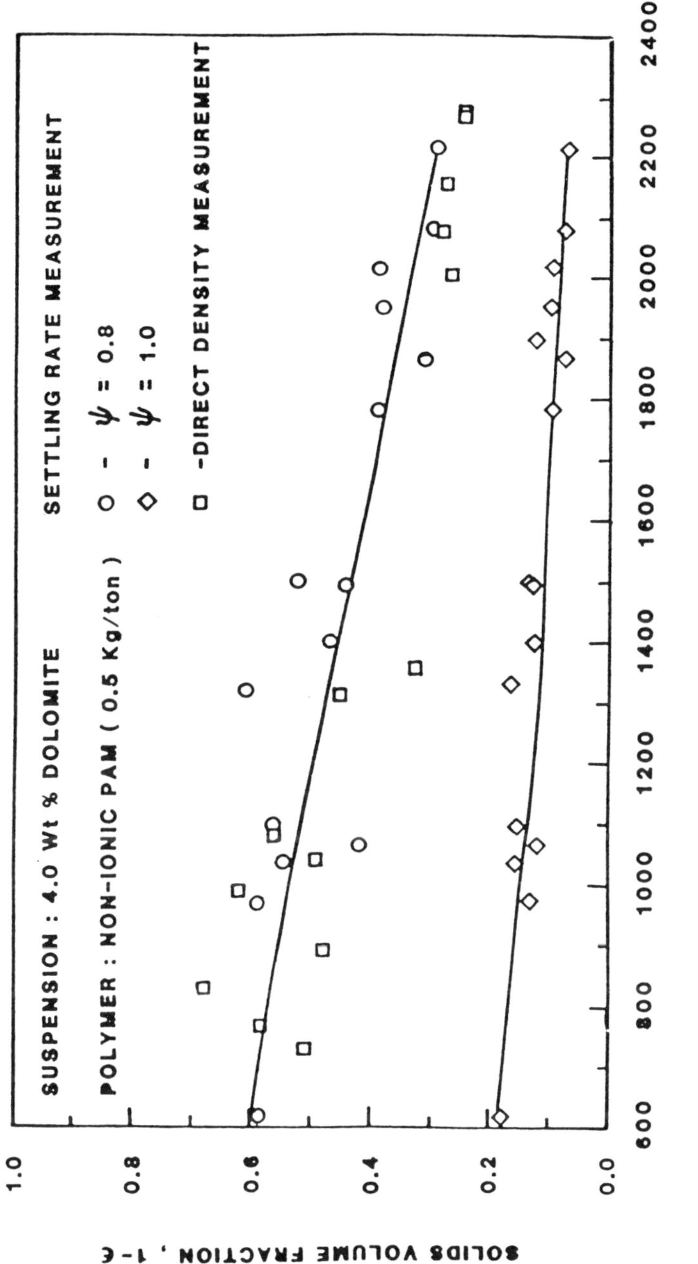

Figure 3. Effect of shape on calculated floc density of dolomite.

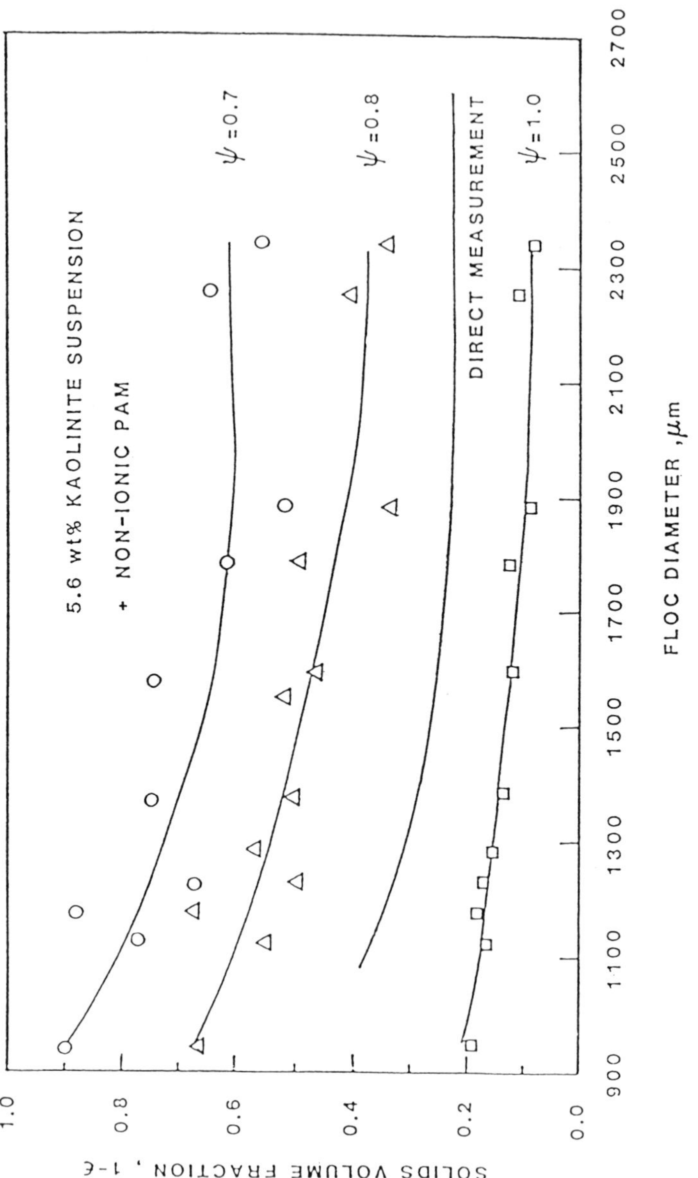

Figure 4. Effect of shape on calculated floc density of Kaolinite.

  Characterization Techniques for the Solid-Solution Interface

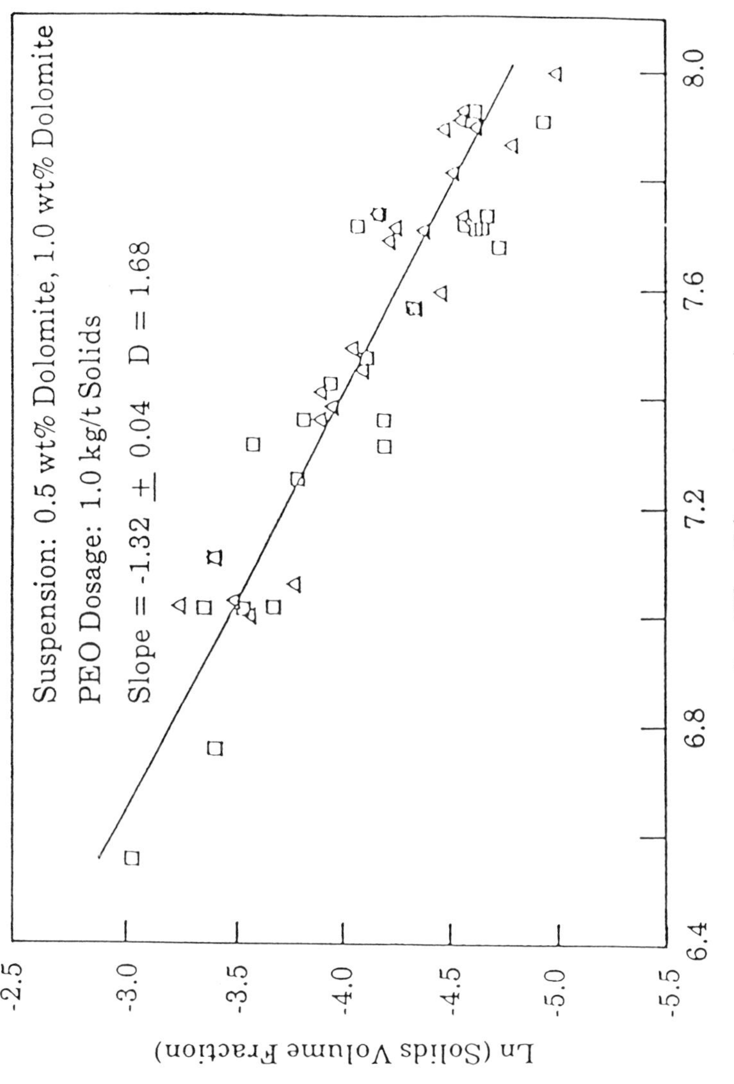

Figure 5. Relationship between floc diameter and solids volume fraction for a dolomite suspension; effect of solids loading on fractal dimension.

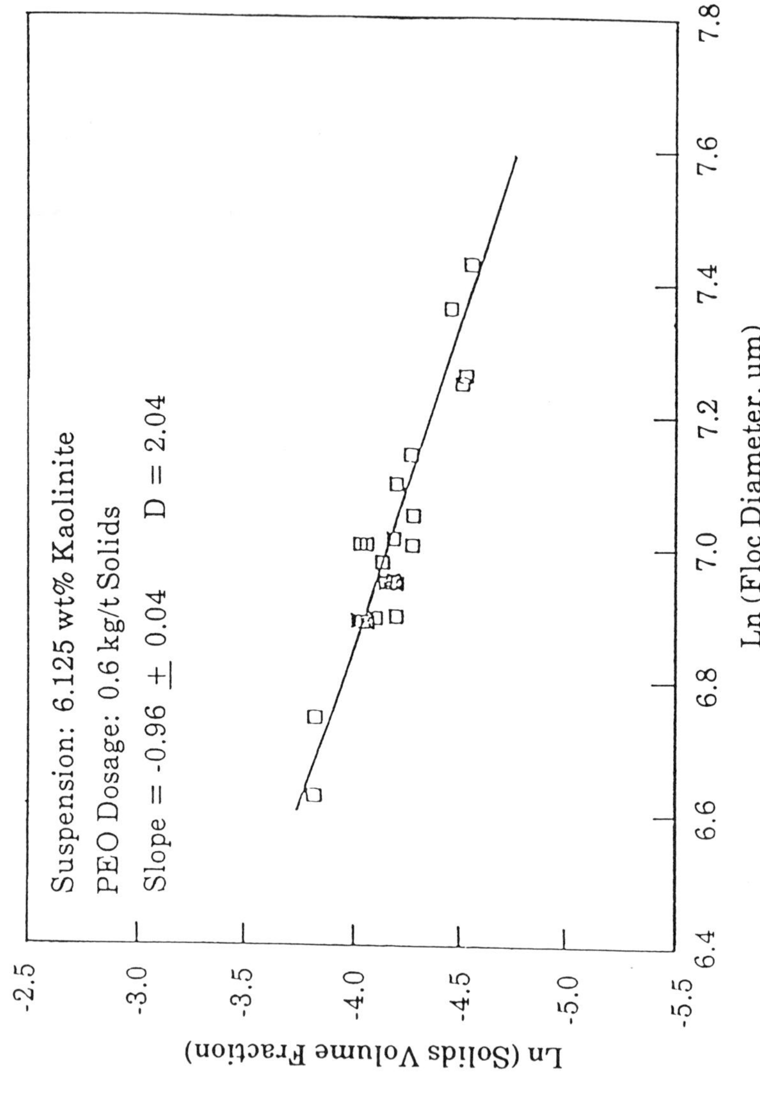

Figure 6. Relationship between floc diameter and solids volume fraction in floc for a Kaolinite suspension.

$$\alpha = \frac{S_f}{a_f} = \frac{\rho_w}{2}\left[\frac{\eta_o D_f}{\rho_w}\right]^{2/3}$$

where η_o = energy dissipation rate per unit volume of suspension, erg/cm^3·sec.

D_f = mean floc diameter obtained after shearing, cm

ρ_w = density of suspension medium, g/cm^3

a_f = cross sectional area of floc, cm^2

Procedure to measure shear strength of flocs

Flocculation tests were conducted using a fully baffled (0.85 cm), 7.6 cm I.D., 14 cm tall and 500 cc standard reaction vessel obtained from the Fisher Scientific Company. It was fitted with a 2.85 cm diameter, six bladed stainless steel turbine impeller for all the flocculation experiments. The power input was determined by a controller (G.K. Heller Corp.) with a torque and velocity readouts.

Illustration

Using dolomite as a substrate binding strength experiments were carried out by Moudgil and Vasudevan using polymers of different types, molecular weights and dosages [12]. It can be seen from the results presented in Table 3 that the binding strength reaches a maximum with an increase in the polymer dosage and then decreases as the dosage is increased further. This phenomena probably results from optimum surface coverage. Furthermore, it can be seen that the binding strength obtained using a high molecular weight polyacrylamide was almost the same as that with the low molecular weight polyacrylamide. However the optimum polymer dosage is 2.5 times greater for the lower molecular weight fraction.

The effect of pH on floc binding strength is illustrated in Figure 7. In these experiments Al_2O_3 was used as a substrate and a high molecular weight anionic polyacrylamide was employed as a flocculant . The binding strength for a given polymer dosage increases as the shearing speed is increased at both pH levels, however the flocs at pH 4 have a substantially lower binding strength than at 7.7. This may be explained by the fact that at higher pH the primary particles are closer to each other due to a lower zeta potential resulting in formation of more compact flocs.

SUMMARY

Presence of soft agglomerates (flocs) of suitable properties such as size, distribution, density and shear strength can be effective in preventing particle segregation in multi component systems and lead to denser green bodies and faster

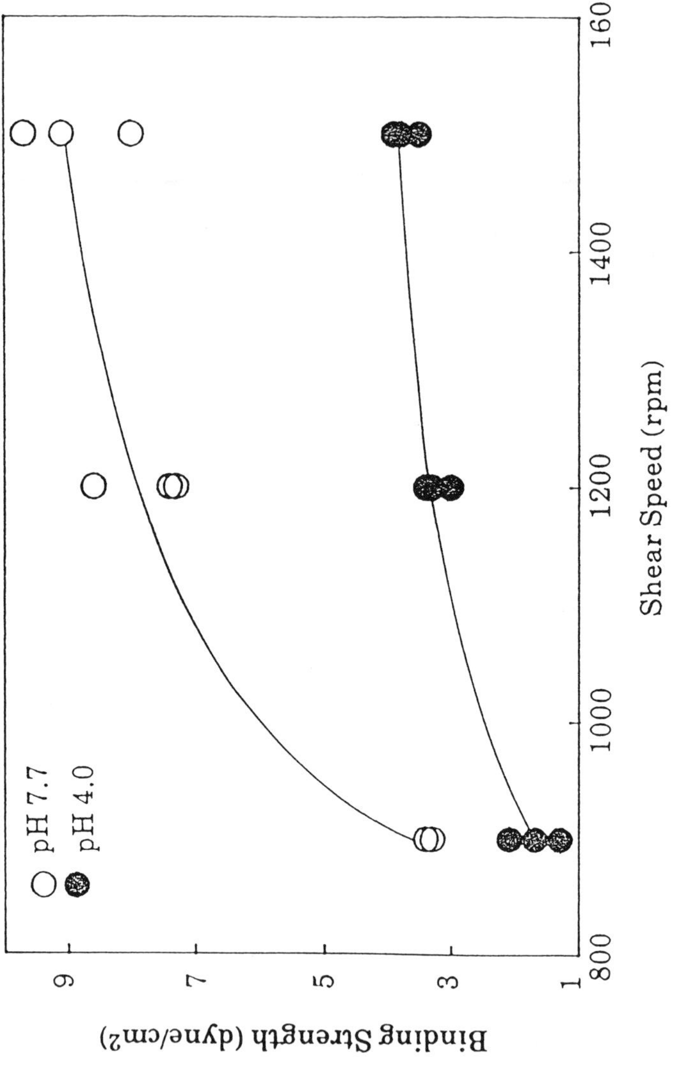

Figure 7. Effect of pH upon floc binding strength.

 Characterization Techniques for the Solid-Solution Interface

casting rates in colloidal ceramic processing. Past characterization attempts to
evaluate the various properties of flocs have been limited. In this study techniques
to characterize size, distribution, density and shear strength of soft agglomerates
have been established. Attempts are underway to evaluate deformability of flocs by
rheological measurements.

Polymer	Polymer Dosage kg/t solids	Shearing Conditions		Binding Strength, α dynes/cm^2
		Time, min	Speed, rpm	
PAA-15*	1.00	0.5	1100	34
	1.25	0.5	1100	46
	2.50	0.5	1100	25
SF-27**	0.25	2.0	1200	25
	0.50	2.0	1200	44
	1.00	2.0	1200	28

* Molecular weight = 3.6million,   ** Molecular weight = 15 million

Table 3. Effect of polymer molecular weight and dosage on binding strength.

ACKNOWLEDGMENTS

The authors acknowledge partial financial support of this work by the
National Science Foundation (MSS # 8821815) and the Mineral resource Research
Center, UF.

REFERENCES

1.   Velamakani, B.V., and Lange, F.F., *J. Am. Ceram. Soc.,* **74** (1), 166-71
        (1991).
2.   Sutherland, D.N., *J. Colloid and Interface Sci.,* **25**, 373-380 (1967).
3.   Koglin, B., *Powder Technol.,* **17**, 219-227 (1977).
4.   Klimpel, R.C., Dirican, C., and Hogg, R., *Particulate Sci. Technol.,* **4**,
        45-59 (1986).

5.   Farrow, J.B. and Warren, L.J., "The Measurement of Floc Density-- Floc Size Distribution", in Flocculation and Dewatering, pp. 153-165. Eds. B.M. Moudgil and B.J. Scheiner, Engineering Foundation, N.Y., 1989.
6.   Dirican, C., "The Structure and Growth of Aggregates in Flocculation", M.S. Thesis, The Pennsylvania State University, 1981.
7.   Tambo, N., and Watanabe, V., *Water Research*, **13**, 409-419 (1979).
8.   Goodarz-Nia, I., "Computer Simulation Studies of Floc Structures," Ph.D. Thesis, University of London, 1972.
9.   Vold, M. J., *J. Colloid and Interface Sci .*,**18**, 684-695 (1963).
10.   Matsumo, K., and Mori, Y., *J. Chem. Engg.(Japan)*, **8**, 143-152 (1975).
11.   Hogg, R., Klimpel, R. C., and Ray, D.T., "Structural Aspects of Floc Formation and Growth", in Flocculation, Sedimentation and Consolidation, pp 217-227. Eds. B.M. Moudgil and P. Somasundaran, Engineering Foundation, N.Y., 1985.
12.   Moudgil, B.M. and Vasudevan, T.V., *Minerals and Metallurgical Processing*, **3**, 142-145 (1989).
13.   Klimpel, R.C. and Hogg, R., *J. colloid and Interface Sci.*, **113**, 121-131 (1986).
14.   Hogg, R., Klimpel, R. C., and Ray D.T., *Minerals and Metallurgical Processing*, **1**, 108-114 (1987).
15.   Matsumo, K., and Suganuma, A., *Chemical Engineering, Sci.*,**32**, 445 (1977).
16.   Meakin, P., *J. Colloid and Interface Sci.*, **102**, 491-504 (1984).
17.   Glasgow, L.A., and Hsu, J. P., *AIChE Journal*, **28**, 779-782 (1982).

     Characterization Techniques for the Solid-Solution Interface

ELECTROSTATIC AND STERIC EFFECTS IN Si_3N_4 BASED SUSPENSIONS

C. Galassi, and E. Roncari
IRTEC - Research Institute for Ceramics Technology
National Research Council
Faenza, Italy

ABSTRACT

Al_2O_3 and Y_2O_3 were added to Si_3N_4 using mechanical mixing and coprecipitation routes. Zetapotential, sedimentation behavior and complex, elastic, and loss moduli were determined in the aqueous suspensions of the resulting powders as a function of pH or the concentration of a graft copolymer used as a strong deflocculant.

INTRODUCTION

The role of colloidal methods for treating and consolidating ceramic powders has been emphasized in terms of improving the reliability of ceramics.(1) Many of the problems connected with powder processing arise from the production of microstructural heterogeneities and the non-uniform phase distributions either associated with the powder itself or developed during treatment and consolidation into a shape. Colloidal methods can bring about remarkable improvements in the microstructure. This has highlighted the importance of characterization methods of powder surface, solid-solution interaction, and adsorption of polymers. Moreover, potential effectiveness of rheological characterization for studying slurry properties was recently investigated.

Typical systems are high solids loading multiphase slurries with remarkable differences in density. They must have properties capable of producing an uniform, cold consolidated dry state so as to encourage the formation of high density, well packed green bodies which, after sintering, generate a product with optimum microstructure. Silicon nitride based materials are well known for their mechanical and physico-chemical properties and as one of the most promising candidates for structural applications. The high degree of covalent bonding in silicon nitride and the addition of sintering aids allow densification in reliable conditions through the promotion of a liquid phase.(2,3) The effect of the addition of the sintering aids through a colloidal route on the densification behavior and on the final properties of the material has been reported elsewere.(4) The more homogeneous distribution modifies the intergranular phase and improves the flexural strength by about 200 MPa.(5)

Many studies have demonstrated the possibility of producing high solids aqueous suspensions but the control of the parameters governing cold consolidation is still a matter for discussion.(6) The control of the dispersion plays a decisive role provided that well dispersed, weakly flocculated or flocculated systems show optimum performance in the different steps of the production process. The aim of this study is to characterize the dispersion state of the two powder mixtures produced with mechanical mixing and through coprecipitation of sintering aids with the silicon nitride powder, by means of electrostatic and steric mechanisms.

MATERIALS AND METHODS

The Si_3N_4 powder (Starck LC12SX) has a specific surface area of 19 m^2/g and an oxygen content of 2.0 wt%. Two mixing routes were used for the final batch composition of 89 wt% Si_3N_4 8 wt% Y_2O_3 3 wt% Al_2O_3. The mechanical mixing was in an aqueous medium with Al_2O_3 (ALCOA A16) and Y_2O_3 (Hoechst) in a ball mill with Si_3N_4 mixing media followed by freeze drying and sieving (M powder). The other mixing route was the coprecipitation of hydroxides on the Si_3N_4 powder from the solution of the nitrates followed by freeze drying and calcination(4) (C powder).

The additives were coprecipitated in the same conditions (AC powder) and mechanically mixed (AM powder) in order to measure their surface area, morphology and zeta potential. Specific surface area was determined using BET method[1], agglomerate size by X-ray transmittance[2], crystalline phase content by X-ray diffraction[3], and morphology by SEM analysis[4]. Aqueous slurries of proper solid content were prepared in order to study sedimentation and rheological behaviour related to pH modification in the basic range or to the addition of a graft copolymer[5].

The electrophoretic mobility of the starting Si_3N_4, C, M, AC, and AM powder was measured as a function of pH in the range 3 to 11 using the Rank Brothers Mark II microelectrophoresis apparatus. Powders were ultrasonically dispersed into a 10^{-3} M KCl solution to maintain a constant ionic strength. The pH was adjusted with NaOH or HCl. Diluted suspensions ($\varnothing \sim 10^{-5}$) were introduced into a capillary cell which was maintained at 25°C. Each mobility value was

---

[1]Sorpty 1750 - Carlo Erba Strumentazioni
[2]Sedigraph 5000 ET - Micormeritics
[3]Giegerflex - Rigaku
[4]Cambridge S360
[5]Atlox 4913 - ICI

       Characterization Techniques for the Solid-Solution Interface

calculated from the average of 40 velocity measurements The zeta potential was obtained according to the Smoluchowski equation.

To evaluate the dispersion state, the sedimentation experiments were performed with 10 vol% suspensions of M and C powders at pH 9-11 adjusted with NH_4OH and after addition of different amounts of the deflocculant. The suspension was poured into a 25 ml measuring cylinder. The sedimentation height of the cloudy layer was recorded as a function of settling time.

The rheological characterization of concentrated (55 wt% equal to 27 vol%) suspensions prepared by using fast stirred ball milling was performed on a shear controlled coaxial cylinder viscometer[1] in order to record the flow curves and measure the viscoelastic properties of the suspension, to which the graft copolymer was added, through strain sweep tests at a frequency of 0.1 Hz.

RESULTS AND DISCUSSION

Powder characterization

Figure 1 shows the morphology of the (a) mechanically, (b) chemically, and (c) mixed batches and as coprecipitated additives, whose specific surface areas are 19, 41 and 192 m^2/g, respectively. The X-ray diffraction patterns of the C powder show only the Si_3N_4 peaks, while the additives are clearly detectable in the M powder. The coprecipitated additives are amorphous, being primary particles in the size range of 50-100 nm.

The zeta potential of the as received Si_3N_4, M, C, AM, and AC powders are reported in Figure 2. The values measured for the as received powder are in agreement with the values reported in literature(7,8) and obtained with different experimental methods.

The behaviour of the mixed powders, M and C, is very similar to that of the starting powder in the acidic range, while the additives play a major role on the basic side. For all powders the isoelectric point is at about pH 4. The zeta-potential increases with a trend similar to the as received powder for the M composition, and to the chemically mixture of the additives for the C one. This highlights the role of the nanometric precipitates on the Si_3N_4 particle surface in relation to the trivalent counter ions present in the solution at low pH. The solubility equilibria of the hydrolyzed oxides is:

$$Me(OH)_3 = Me^{3+} + 3(OH)^- \text{ with } Me = Al \text{ or } Y$$

with partial dissociation occuring in the acidic range. At pH > 8.5 the concentration of ions is very low.(9,10) The additives are almost undissociated. When they cover the surface it behaves as an oxide-like surface.

---

[1]Bohlin VOR

Figure 1. SEM morphologies of: (a) mechanically mixed Si_3N_4 and additives, (b) chemically mixed Si_3N_4 and additives, and (c) coprecipitated additives.

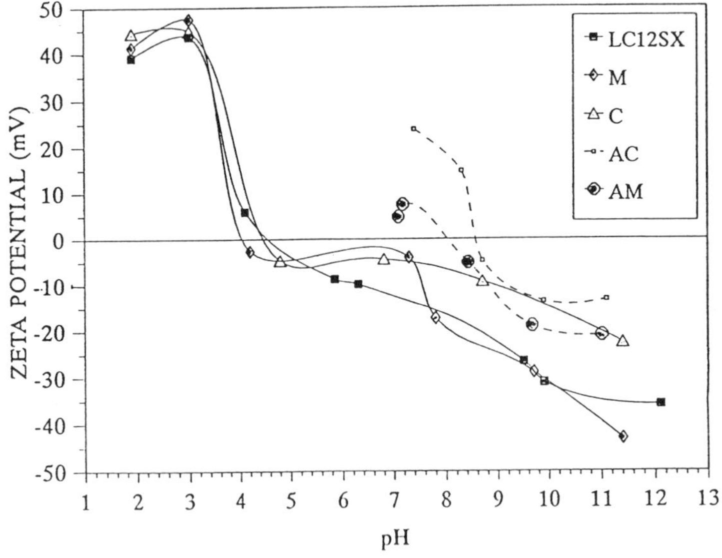

Figure 2. Zeta potential of Si_3N_4 suspensions vs. pH.

Settling experiments and adsorption

The zeta-potential results show that electrostatic stabilization of the powders is hardly a good means of stabilizing high volume suspensions because of the low value of zeta-potential even at high pH (max 40 mV for M powder) while it can not be proposed for the C powder. A graft copolymer was employed to check its effectiveness as steric stabilizer. The set of settling experiments is reported in Table 1. The data reveals the effectiveness of the pH modification on the stabilization of 1O vol% suspensions as expected from zeta-potential measurements on the M powder. A pH adjustment does not affect the behaviour of the C powder. In fact, the M powder flocculates promptly in a high volume sediment with a clear supernatant at pH 5, while a cloudy suspension with a very low volume sediment is present in the 9-11 pH range. The addition of the deflocculant results in a stabilizing effect more pronounced in the higher pH value (10.3) for the M powder while acceptable results are obtained for the C powder only at pH 7.3. The deflocculant promotes flocculation of the M powder at pH 9 while it is prevented at pH 10.3. This provides evidence of a strong effect of pH on the behaviour of the deflocculant.

The graft copolymer used works through an anchoring group and stabilizing group which is soluble in the dispersing medium and must be in a good solvent condition. On the C powder, where the electrostatic mechanism seems not to work, the addition of the deflocculant produces a stabilizing effect (albeit weak) only at a lower pH value (7.3) without the addition of ammonia. Therefore, while in the case of M powder there is competition between the two effects though the major role is played by the electrostatic effect, in the C powder only the steric effect is active and, besides the quantity of deflocculant, the solvency of the medium plays a critical role.

Rheological characterization

The rheological characterization was performed on 27 vol% suspensions at increasing concentrations of deflocculant. The average values of the complex (G^*), storage (G') and loss (G'') modulus in the linear region were recorded and plotted against the deflocculant concentration in Figure 3.(11,12) For both powders the complex moduli coincides with the storage modulus, thus highlighting the major role of the elastic compared to the viscous component.

The adsorption of the deflocculant produces a continuous increase of the moduli in the M suspensions that evidences a gradual increase in the flocculation with the increase in deflocculant concentration. The C suspensions, although at much lower solids concentration, show values of the moduli of the order of magnitude higher than M powder at low deflocculant concentration.

This, together with a progressive reduction of the strain amplitude at which the non-linear response arises, evidences a strong enhancement of flocculation, with a "harder" structure. After the sharp rise the moduli decrease with further

	Test	pH	Deflocc.wt %	Sediment height 24 h	Sediment height 100 h
M	1	5.0	0	8.4	6.9
	2	9.3	0	0.6	1.5
	3	10.3*	0	11.0	11.0
	4	11.0	0	1.7	2.4
	5	7.0	0.50	7.5	7.5
	6	9.0	0.18	11.3	10.0
	7	9.0	0.50	10.5	9.5
	8	10.3	0.17	0	7.5
	9	10.3	0.50	1.9	2.4
	10	10.3*	0.50	11.2	10.6
	11	10.3	0.66	0.6	0.7
	12	10.3	0.83	0.5	0.5
C	1	9.0	0	10	10.0
	2	10.3	0	10.8	10.8
	3	11.0	0	10.9	11.5
	4	7.3	0.33	6.2	6.6
	5	7.3	0.50	7.6	7.8
	6	7.3	0.66	7.6	7.8
	7	10.3	0.33	8.7	8.8
	8	10.3	0.50	12.0	11.6
	9	10.3	0.66	10.0	10.0

* 27 vol. % solid

Table 1. Settling experiments (10 vol% solid).

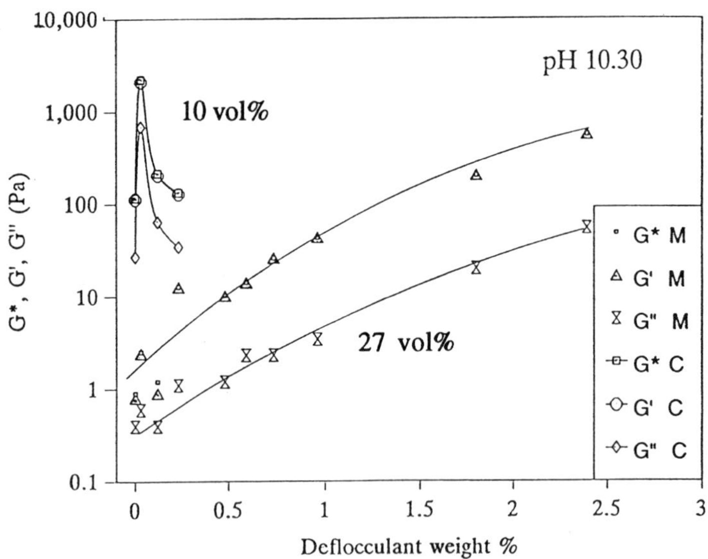

Figure 3. Elastic moduli versus deflocculant concentration for M and C suspensions.

  Characterization Techniques for the Solid-Solution Interface

addition of deflocculant, thus indicating a partial restabilization, probably due to the reduction of the floccule size.

CONCLUSIONS

The characterization of silicon nitride suspensions, either based on mechanical or chemical mixing of the sintering aides was performed. Adjustment of suspension pH produces stabilization of only relatively high solids loading suspensions of mechanically mixed powder. The effects of the addition of a graft copolymer were tested with sedimentation experiments and measurement of the complex, elastic, and loss moduli in the linear viscoelastic region. Enhanced flocculation was detected in both cases, but more pronounced in the chemically mixed powder based suspensions. It is due probably to the poor solvent conditions of the commercial graft copolymer in water. Viscoelastic characterization seems to be a powerful technique to characterize the dispersion state and will be more deeply applied to the ceramic systems.

REFERENCES

1. F.F. Lange," Approach to reliable powder processing," pp.1069-83 in Ceram. Trans. "Ceramic Powder Science II vol. B" Ed. G.L. Messing et. al. The Am. Ceram. Soc., OH, 1988.
2. G. Wotting, and G.Ziegler, "Powder characteristics and sintering behaviour of Si_3N_4," Interceram **2**, 25-32(1986).
3. J.S. Kim, H. Schubert, and G. Petzov, "Sintering of Si_3N_4 with Y_2O_3 and Al_2O_3 added by coprecipitation," Europ.Ceram.Soc., **5**, 311-19(1989).
4. T.M. Shaw, and B.A. Pethica, "Preparation and sintering of homogeneous Silicon Nitride green compacts," J.Am.Ceram.Soc., **69**[2], 88-93(1986).
5. C. GAlassi, V. Biasini, and A. Bellosi, "Effects of powder characteristics and mixing processes on microstructure and properties of Silicon Nitride," in preparation
6. I.A. Aksay, "Microstructure Control Through Colloidal Consolidation," pp.94-104 in *Forming of Ceramics,* Ed. J.A.Mangels,and G.L.Messing, Advances in Ceramics vol 9, The Am.Ceram.Soc., OH, 1984.
7. M.J. Hoffmann, A.Nagel, P.Greil, and G.Petzow, "Slip casting of SiC-whisker-reinforced Si_3N_4," J. Am. Ceram. Soc., **72**[5], 765-69(1989).
8. P.Greil, A.Nagel, H.Stadelmann, and G.Petzow, "Review: Colloidal processing of Silicon Nitride Ceramics," pp.319-29 in Proc. Ceramic Materials and Component for Engines. Ed. V.J.Tennery , The Am. Ceram. Soc., OH, 1989.
9. E.F. Adams, *Slip cast ceramics, Part IV - High temperature oxides,* Academic Press, NY, 1971.

10. R.D. Shannon, "Revised Effective Ionic radiu and Systematic Studies of Interatomic distances in Halides and Chalcogenides," Acta Crystallog. Sect. A, **32**, 571(1976).
11. Th.F. Tadros, "Physical Stability of Suspension Concentrates," Adv. Colloid. Interface Sci., **12**[2-3], 141(1980).
12. Th.F. Tadros, "Control of the properties of suspensions," Colloids and Surfaces **18**, 137-173(1986).

 Characterization Techniques for the Solid-Solution Interface

CARBOXYLIC ACIDS AS DISPERSANTS FOR ALUMINA SLURRIES

P.Hidber, Th.Graule and L.J.Gauckler
Nichtmetallische Werkstoffe
ETH Zürich
CH-8092 Zürich

ABSTRACT

The dispersing effect of different carboxylic acids on high purity alumina suspensions in aqueous medium was investigated. The contributions of the different carboxylic acid building blocks to their dispersing potential could be characterized by the adsorption behavior of the dispersant on the alumina powder surface and by measurements of the dynamic electrophoretic mobility $\mu_d(\omega)$.
An explanation for the well known good dispersing effect of citric acid for alumina slurries is given.

INTRODUCTION

Stabilization of highly concentrated aqueous ceramic slurries is achieved by
a) electrostatic repulsion forces between the particles due to surface charges either produced by reaction of the particles with the liquid [1] or by adsorbed charged molecules [2]
b) electrosteric hindrance, based on charged surfaces and the hydrostatic overpressure in the regions where long chain polymer molecules, attached to the particle surface, overlap.
Organic molecules of less than 200 g/mol are adsorbed on solid surfaces only if they have at least one functional group [3]. These functional units like carboxyl, hydroxyl or amino groups may substitute surface hydroxyl groups and are able to form complexes between the metal ion of the surface and the specific adsorbing molecule. The adsorption of carboxylic acids [4], sulfonic acids [5] and amines [6] on oxide surfaces have been investigated as well as their influence on the stability of dispersions [7].

The aim of the present study is to characterize the influence of aliphatic and aromatic carboxylic acids on the stability of alumina dispersions by measuring the dynamic electrophoretic amplitude $\mu_d(\omega)$ and the adsorption behavior as a function of pH and additive content.

Characterization Techniques for the Solid-Solution Interface

EXPERIMENTAL

Materials:

High purity alumina (HRA10; >99.99% Al_2O_3; 0.0030% Na_2O) from Showa Denko Corp.,Japan, was used in this investigation. The average particle diameter is 0.5 μm, the BET 10 m^2/g.

L-lactic acid, DL-malic acid, tricarballylic acid, citric acid, 2-hydroxybenzoic acid, 3-hydroxybenzoic acid and 4-hydroxybenzoic acid (99% minimum purity, Fluka Chemie AG, CH-Buchs) were used without further purification. KNO_3 used in adsorption experiments was of p.a. quality (Merck, Darmstadt).

[14]C-labeled citric acid was obtained from Amersham Laboratories, UK.

All solutions were prepared with high purity water (el. resistivity > 20 MΩcm).

Methods:

Electrophoretic mobility measurements:

Suspensions of 15 wt% Al_2O_3 in water with 0.4 wt% dispersant (solids = 100%) were prepared by dispersing the powder with 4 min ultrasonic treatment[*]. ESA potentials[**] were measured at different pH values by titrating the suspension automatically with 1N KOH or 1N HCl. From the ESA potential, the dynamic electrophoretic mobility $\mu_d(\omega)$ is calculated:

$$\mu_d(\omega) = \frac{ESA(\omega)}{c \cdot \Delta\rho \cdot \Phi \cdot G_f}$$

c: sound velocity in the suspension
$\Delta\rho$:   difference in density between particles and liquid
Φ: volume percent of solid
G_f: geometrical factor

Equilibrium adsorption experiments:

Alumina (7 wt%), organic acid solution (usually 1mM), inert electrolyte (usually 0.1M KNO_3) and an aliquot of HNO_3 or KOH were filled into 30ml polypropylene tubes. After an equilibration time of 48h at 25°C, the pH of the suspensions was measured. The concentration of the free organic acid was measured after separation of solid matter by centrifugation either by UV absorption spectroscopy[+] (in case of the aromatic carboxylic acids) or by determination of the radioactivity[++] (in case of the ^{14}C labeled aliphatic carboxylic acids).

---

[*] 600W Vibra-cell ultrasonic desintegrator VC600, Sonics & Materials, USA

[*] Matec ESA system 8050 from Matec Applied Sciences, Hopkinton, USA

[+] Beckman DU 62 Spectrophotometer, Beckman Instruments Intern. S.A., CH-1260 Nyon

[++] Beckman LS1801 Szintillation counter, Beckman Instruments Intern. S.A. CH-1260 Nyon

RESULTS AND DISCUSSION

Citric acid as a dispersant

The dynamic electrophoretic mobilities of slurries with increasing amounts of citric acid are shown in Figure 1:

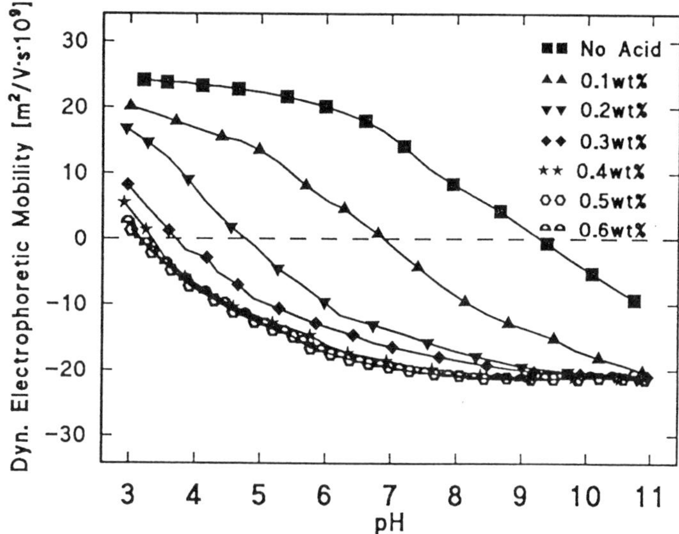

Figure 1: Electrophoretic mobilities of alumina slurries with citric acid additions.

For the pure alumina slurry the results reveal the previously reported behavior [8] for high purity alumina samples. The isoelectric point is at pH 9.2. Contrary to other publications no drop in electrophoretic mobility is observed when reaching the strongly acidic region.

A 0.1 wt% addition of citric acid results in a shift of the isoelectric point (IEP) by 2 pHunits. Further increase of citric acid shifts the IEP further to the acidic region. In the basic region, the electrophoretic mobility saturates at - 22 $10^{-9}m^2$/Vs due to a saturation of the alumina surface by citric acid molecules. This saturation is also observed in adsorption experiments (Figure 2). Over a wide pH range (pH 3-8) constant adsorption is observed for suspensions with 0.1 wt% citric acid. At higher pH values adsorption decreases strongly. The amount of adsorbed citric acid increases linearly up to 0.2 wt% acid. Further addition leads to a gradual saturation. Adsorption isotherms demonstrate that at low pH values nearly complete adsorption is possible, whereas only a fraction of the citric acid is adsorbed in case of higher pH values.

To elucidate this well known dispersant effect of citric acid, we compared the effect of other different carboxylic acids.

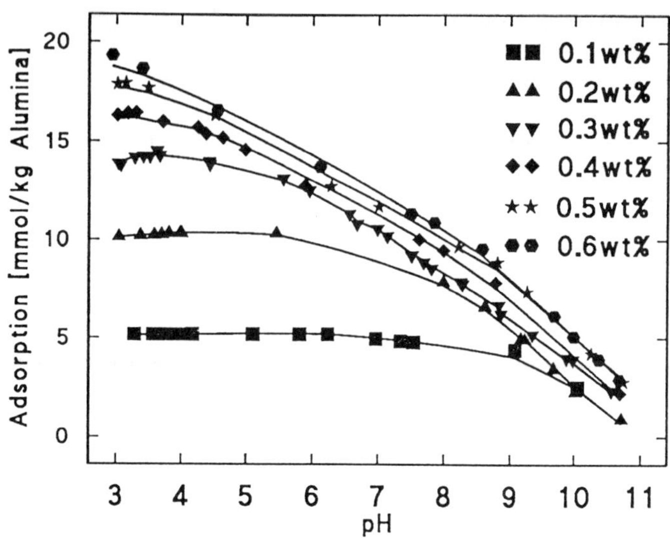

Figure 2: Adsorption behavior of citric acid on alumina.

The importance of charge carriers

The dynamic electrophoretic mobilities of alumina slurries with and without addition of lactic acid, malic acid and citric acid are shown in Figure 3.

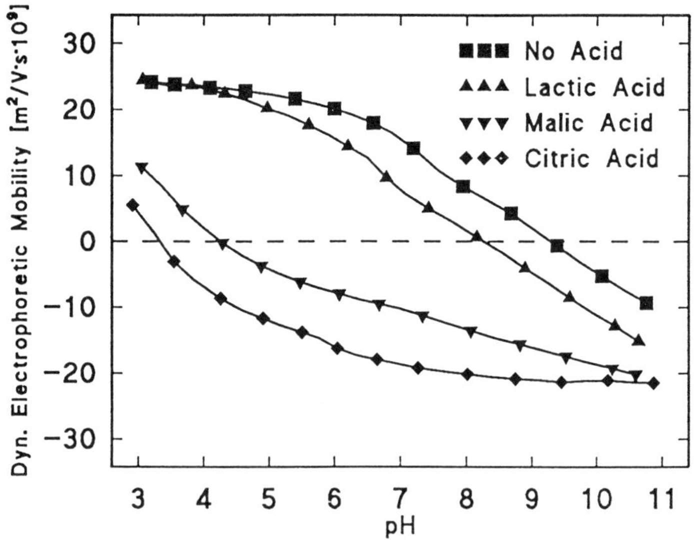

Figure 3: Electrophoretic mobilities of slurries with hydroxycarboxylic acids.

   Characterization Techniques for the Solid-Solution Interface

The importance of the steric arrangement of the functional groups

We have clearly shown the importance of the hydroxyl group on adsorption efficiency and therefore the electrophoretic mobilities. Remains the question, whether the steric arrangement of the functional group might be of influence. This can be clarified by using isomeric hydroxybenzoic acids, as shown in Figure 5.

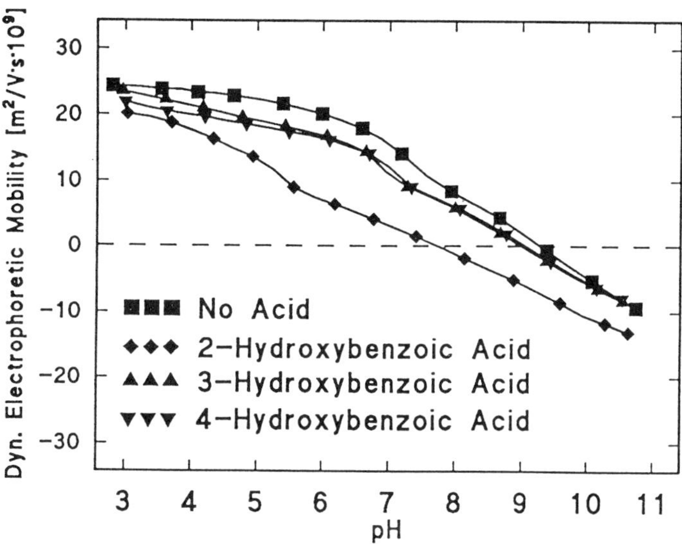

Figure 5: Electrophoretic mobilities of alumina slurries with hydroxybenzoic acids

The 3- and 4-hydroxybenzoic acids show only a slight dispersing effect in the acidic pH range comparing their corresponding mobility curves with a pure alumina slurry. No effect is observed in the alkaline region. The 2-hydroxybenzoic acid shifts the isoelectric point by 1.5 pH units to the acidic region. As it can act as a two-fold ligand, it adsorbs very good on the surface of the alumina powder.

The adsorption measurements (Figure 6) confirm these observations. Whereas 2-hydroxybenzoic acid adsorbs strongly between pH 4 and 8, only slight adsorption at around pH 4 is seen for 3- and 4-hydroxybenzoic acid. This was already observed for γ-Al$_2$O$_3$ [10].

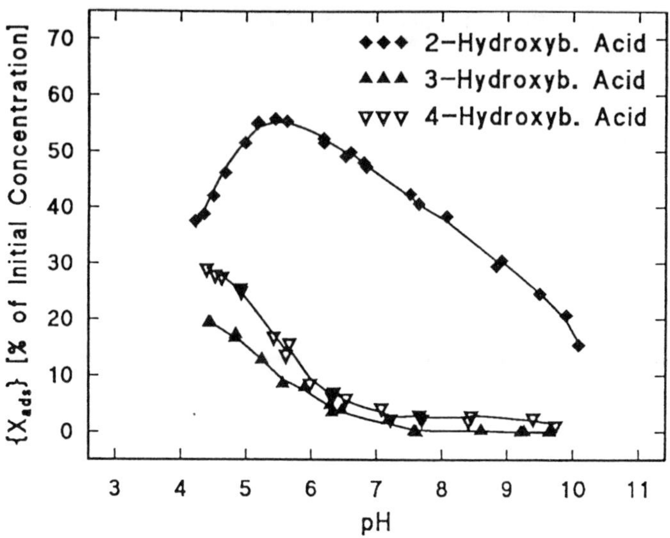

Figure 6: Adsorption behavior of hydroxybenzoic acids on alumina.

We conclude therefore, that the specific geometric arrangement formed by the COOH- and OH- groups in the 2-hydroxybenzoic acid is more advantageous compared to all other steric arrangements investigated in this study.

CONCLUSIONS

The following requirements are advantageous for a carboxylic acid to serve as a dispersing agent in alumina slurries:

- The efficiency of a carboxylic acid as a deflocculant for alumina slurries depends on the amount of charge carrier groups and additional functional groups as well as on the steric arrangement of the different functional groups.
- Increasing the amount of charge carriers per molecule leads directly to higher surface charges on the particles and to larger changes in the electrophoretic mobility curves.
- The hydroxyl group as an additional functional group enhances the ability for an organic molecule to adsorb on alumina especially in the alkaline pH region.
- Hydroxybenzoic acids only act as dispersants if the steric arrangement of the hydroxyl and the carboxyl group allows chelate ring formation (e.g. 2-hydroxybenzoic acid).
- The behavior of the electrophoretic mobilities of alumina slurries with carboxylic acid additions are in good correlation with adsorption results.

   Characterization Techniques for the Solid-Solution Interface

In contrast to lactic acid, which shifts the electrophoretic mobility curve only slightly, the di- and tricarboxylic acids show drastic shifts, correlating strongly with the increasing charge carrier amount per molecule. Thus, even the adsorption is almost similar for malic and citric acid [9], electrophoretic mobility is much more influenced in the case of citric acid.

The importance of adsorption efficiency

The increase of the adsorption efficiency due to the hydroxyl groups of the dispersant can easily be demonstrated by comparing the electrophoretic mobility curves for slurries with tricarballylic acid and citric acid as shown in Figure 4.

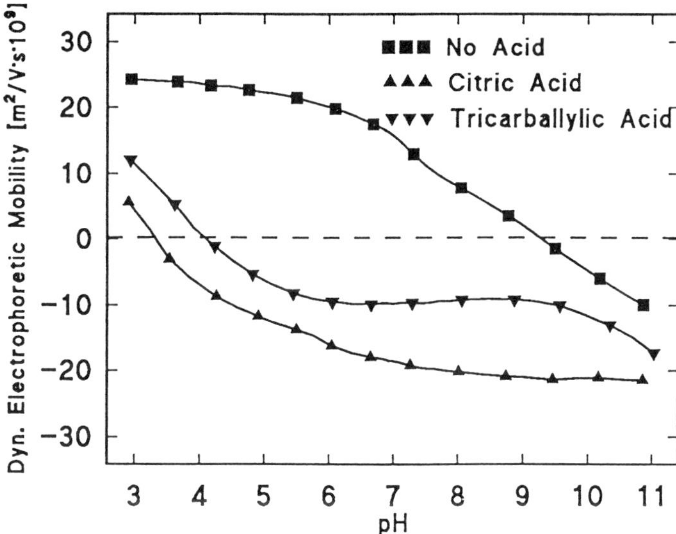

Figure 4: Electrophoretic mobilities of alumina slurries with tricarboxylic acids.

In the neutral and alkaline pH region citric acid is still able to form surface complexes due to its hydroxyl group and therefore adsorbs on the alumina surfaces leading to a drastic electrophoretic mobility change. In contrast to citric acid, the tricarballylic acid does not possess this hydroxyl group being otherwise the identical molecule with the same amount of charge carriers. Tricarballylic acid is thus less adsorbed and less efficient as dispersant compared to citric acid especially in the alkaline region. This proves the importance of the hydroxyl groups for chelate complex formation and better adsorption possibility in the alkaline region.

ACKNOWLEDGEMENTS

We thank the Swiss KWF organization and the Alusuisse-Lonza Services AG for financial support.

REFERENCES

[1]   I.A. Aksay, "Microstructure Control Through Colloidal Consolidation", Advances in Ceramics, Vol. 9, Forming of Ceramics, pp. 94 - 104, edited by J.A. Mangles and G.L. Messing, American Ceramic Society, Columbus, OH, 1984.

[2]   G.A. Parks and P.L. de Bruyn, "The Zero Point of Charge of Oxides", J. Phys. Chem., 66, 967-973 (1962).

[3]   R. Kummert and W. Stumm, "The Surface Complexation of Organic Acids on Hydrous γ-Al$_2$O$_3$", J. Colloid Interface Sci., 75, 377-385 (1980).

[4]   S. Engels, H. Lausch and R. Schwokowski, "Untersuchungen zur Adsorption bi- und polyfunktioneller organischer Säuren an Aluminiumoxid; (Investigations of the adsorption of bi- and polyfunctional organic acids on alumina)", Chem. Techn., 39, 387-391 (1987).

[5]   P. Somasundaran and D.W. Fürstenau, "Mechanism of Alkyl Sulfonate Adsorption at the Alumina-Water Interface", J. Phys. Chem., 70, 90-96 (1966).

[6]   A. Srinivasa Rao, "Electrophoretic Mobility of Alumina, Titania and Their Mixtures in Aqueous Dispersions", Ceram. Int., 14, 71-76 (1988).

[7]   B.V. Kavanagh, A.M. Posner and F.P. Quirk, J. Colloid Interface Sci., 61, 545 (1977).

[8]   D.W. Cannon and R.V. Mann, "Electrokinetic and Surface Charge Characterization of a Commercial Aluminum Oxide Ceramic Powder using the Matec ESA System", CMS Application Note 350, Matec Instruments Inc., Hopkinton, MA.

[9]   P. Hidber, unpublished results.

10]   R. Kummert, "Die Oberflächenkomplexbildung von organischen Säuren mit γ-Aluminiumoxid und ihre Bedeutung für natürliche Gewässer/ (Surface complex formation of organic acids with γ-alumina and their importance for natural waters)", Ph.D. Thesis, Swiss Federal Institute of Technology, Zurich (1979).

SURFACE CHEMISTRY AND RHEOLOGY OF ZrO$_2$ SUSPENSIONS CONTAINING POLYACRYLATE: EFFECTS OF MOLECULAR WEIGHT AND ZrO$_2$ SURFACE AREA

Y.K. Leong, T.W. Healy and D.V. Boger
Advanced Mineral Products Research Centre, The University of Melbourne, Parkville 3052, Australia

Abstract:

The effect of polyacrylate molecular weight on the yield stress-pH behaviour of Z-Tech ZrO$_2$ (15.1m^2/g) suspensions was evaluated. Also included in the evaluation was the effect of citric acid. The molecular weights used were 2103, 4810, 13600 and 52000 for sodium polyacrylate and 2000 and 750000 for polyacrylic acid (PAA). In general, there is a lowering of both the pH of maximum yield stress and the magnitude of maximum yield stress (τ_{ymax}) as a result of polyacrylate and citric acid addition. The τ_{ymax} reduction is most significant for the small citric acid molecule and, low molecular weight polyacrylate. Significant bridging flocculation may be responsible for the less effective reduction in the τ_{ymax} with high molecular weight polyacrylate. The decrease in the pH of τ_{ymax} or isoelectric point (IEP) with increasing polyacrylate concentration appears to be similar for all molecular weight polyacrylate. A lower surface area MEL ZrO$_2$ (3.0m^2/g) exhibits a similar decrease in the pH of τ_{ymax} and IEP with increasing concentration of polyacrylic acid (Mw=2000). For a given IEP difference (IEP$_{no\,PAA}$-IEP$_{PAA}$), both MEL and Z-Tech ZrO$_2$ required the same surface coverage of polyacrylic acid (gPAA/m^{2}ZrO$_2$). MEL ZrO$_2$, being five times smaller in surface area, requires five times less polyacrylic acid to achieve the same surface coverage. A low viscosity MEL ZrO$_2$ suspension with a solids volume fraction approaching 0.6 has been prepared with sodium polyacrylate of molecular weight 2103.

Introduction:

The effect of molecular weight (Mw) of adsorbing polyelectrolyte on the rheology of concentrated ceramic suspensions is not well understood as indicated by the lack of literature in this area. The recent work by Cesarano and Aksay [1] illustrated the importance of polyacrylic acid Mw, ranging from 1800 to 50000, on the viscosity of concentrated α-alumina suspensions. The suspension viscosity obtained for a range of polyacrylic acid (PAA) Mw and concentrations. was evaluated at a pH of 9.0. For each PAA Mw, a minimum suspension viscosity was observed to occur over a range of PAA concentration. For low Mw PAA the onset of minimum suspension viscosity occured at a much lower PAA concentration and the viscosity rise began at a much higher concentration. The viscosity rise was very pronounced for the high Mw PAA. The minimum suspension viscosity is slightly higher for the high Mw PAA.

There are a number of reports on the effects of polyacrylate (PA) on the ceramic suspension stability and rheology [2-5]. Electrokinetic measurements show that the isoelectric point (IEP) of ZrO_2 and Al_2O_3 is more acidic in the presence of PA [2,4]. At the IEP, ZrO_2 suspensions exhibit a maximum in yield stress [6] and viscosity [2,6]. Rheological measurements show that concentrated ZrO_2 suspensions have a lower pH of maximum yield stress (τ_{ymax}) and maximum viscosity in the presence of PA [3,7]. Leong et al [7,8] have shown that the IEP and the pH of τ_{ymax} have the same value for any given concentrations of PA. It was also reported [3,7,8] that the τ_{ymax} is significantly reduced by PA, i.e., from about 400Pa (no PA) to 100Pa at 1.0dwb% PA ($gPA/100gZrO_2$). The reduction in τ_{ymax} was attributed to a physical and steric mechanism of the adsorbed PA thereby keeping the interacting ZrO_2 particles further apart and reducing the van der Waals attraction. Ionic strength did not affect the pH at τ_{ymax} and magnitude of τ_{ymax} for ZrO_2 containing PA [8]. This ionic strength study revealed that the adsorbed PA is lying flat on the ZrO_2 surface at the IEP.

Yield stress (τ_y) is a measure of the strength of the network structure formed by flocculated particles in the suspension. The magnitude of τ_y is dependent upon the strength of each particle-particle attraction or bond and the number of particle-particle bonds per unit area. The strength of each particle-particle attraction is governed by the van der Waals equation where the magnitude of the attractive force depends on particle size, material and suspending medium Hamaker constant, and the distance of separation between the interacting particles. For a given material, the van der Waals attraction is weaker for smaller particles and for greater distance of separation between the

 Characterization Techniques for the Solid-Solution Interface

particles. The distance of separation between the particles may be increased by adsorbing additives. The objective of this study is to evaluate the effects of molecular weight of adsorbing additives on the τ_y-pH behaviour of ZrO_2 suspension.

Materials and Methods:

The two ZrO_2 powders used in this study were supplied by Z-Tech (ICI Advanced Ceramics Australia) and Magnesium Elektron Inc. (MEL). The BET surface area is 15.1 and $3.0m^2/g$ for Z-Tech and MEL powders respectively. The particle size distributions measured with the Leeds and Northrop Microtrac particle size analyser for the Z-Tech and MEL zirconia are 0.58 and 1.69μm respectively for D_{90}, 0.25 and 0.62μm for D_{50}, and 0.13 and 0.34μm for D_{10}. The density of both ZrO_2 is approximately $6000kg/m^3$. The polyacrylate used were two polyacrylic acids of molecular weight 2000 and 750,000 supplied by Aldrich Chemical company, and four sodium polyacrylate (NaPA) of molecular weight 2103, 4810, 13600 and 52000 supplied by Allied Colloids Ltd. The degree of neutralisation as sodium salt for the NaPA is 88.2, 94.7, 97.5 and 78.6% for Mw of 2103, 4810, 13600 and 52000 respectively. Gel permeation chromatography calibrated with a NaPA standard was used to determine the molecular weight distribution of the NaPA. The properties of the NaPA are listed in Table 1.

Table 1: Properties of sodium polyacrylates.

Polyacrylate	% Activity	% Neutral- isation as Na salt	Weight ave. mol. wt., Mw	Number ave., Mn	Peak Mol. wt., Mp
DP6-3578A	39.7	88.2	2103	1708	1715
DP-2695	39.8	94.7	4810	2680	3090
Versicol EN7	23.7	97.5	13600	5650	10100
Versicol EN9	28.0	78.6	52000	19800	46900

In the preparation of the suspensions, appropriate amounts of the polyacrylate solution, milli-Q water and ZrO_2 were placed in a 50ml container and then sonicated for 1min with a 0.75in horn of Sonifier B-30. The vibrational output is 20kHz and the maximum power output is 350W. The sonifier was operated between 50 to 60% of the maximum power output. The amount of suspension prepared each time is no more than 100g. The prepared suspensions were rested for at least a day prior to any measurements.

Concentrated nitric acid (1M to 15.8M) and potassium hydroxide (3M) were used to change the pH of the suspensions so as to minimise the extent of dilution of the suspensions. Localised flocculation in the suspension was observed to occur at the vicinity of the acid or alkaline droplets which were redispersed by sonication with the high intensity horn. After each pH change the suspension was rested for at least 2hrs and then stirred vigorously with a spatula before conducting the yield stress measurement. An Orion 720A pH meter was employed for the pH measurement. The yield stress was measured with a vane rheometer [9,10].

The surface properties such as surface charge, mobility and IEP, of the ZrO_2 were determined with a Matec MBS-8000 electroacoustic instrument [11]. The advantage of this instrument is that the surface properties of ZrO_2 can be determined at relatively high concentrations of up to 5 vol% solids.

Results and Discussion:

The yield stress-pH behaviour for a 57wt% Z-Tech ZrO_2 treated with NaPA of Mw 2103, 4810, 13600 and 52000 is shown in Figures 1(a), (b), (c) and (d) respectively. The results in Figure 1(e) and (f) are for polyacrylic acid (PAA) of Mw 750000 and citric acid (Mw 192) respectively. The shift in the τ_y-pH profile and the pH of τ_{ymax} to a more acidic pH are clearly evident for all additives. More interestingly, the shift in the pH of τ_{ymax} is accompanied by quite a dramatic reduction in the τ_{ymax} for some additives, in particular, citric acid and low Mw NaPA. Two factors determine the extent of τ_{ymax} reduction and these are the additive concentration and Mw. In general, the reduction in τ_{ymax} increases with increasing additive concentrations.

The reduction in τ_{ymax} by adsorbing additives is attributed to a physical and steric mechanism which keeps the interacting particles further apart [3,8]. At low additive concentration, there is not enough additive molecules to completely cover the surface of the ZrO_2. As a result, the particle network structure of the

     Characterization Techniques for the Solid-Solution Interface

Figure 1: Yield stress-pH behaviour of 57wt% ZrO_2 suspensions treated with sodium polyacrylate of molecular weight a)2103, b)4810, c)13600 and d)52000, polyacrylic acid of Mw e)750000 and f) citric acid.

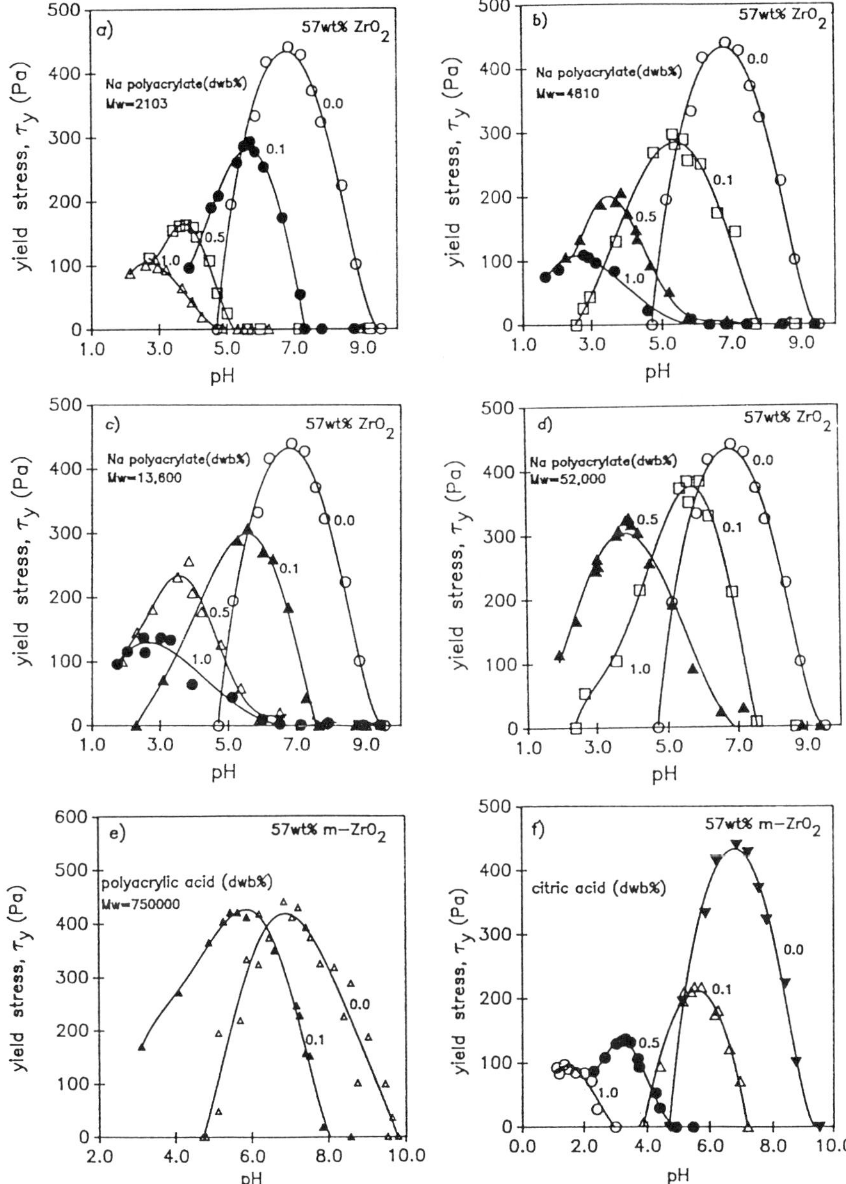

flocculated suspension is formed by mixed surface interactions i.e., between uncoated surfaces, between additive coated and uncoated surfaces, and between coated surfaces. It is obvious that the van der Waals attraction is smaller for interaction between coated surfaces as the surfaces are separated by distance at least equal to the thickness of two adsorbed additive layers. Consequently, the τ_y of the particle network structure is smaller as the proportion of coated surface interactions increases. This explains the smaller τ_{ymax} at higher additive concentrations.

The effect of additive Mw on τ_{ymax} is shown in Figure 2. The plot in Figure 2 is in terms of τ_{ymax} as a function of equivalent polyacrylic acid Mw at fixed additive concentrations of 0.1, 0.5 and 1.0dwb%. The NaPA of Mw 2103, 4810, 13600 and 52000 has an equivalent polyacrylic acid Mw (and n=number of acrylic acid repeating unit) of 1656(n=23), 3730(51.8), 10476(145.5) and 41929(582.3) respectively. From the NaPA Mw and the degree of neutralisation as sodium salt in Table 1, the number of acrylic acid repeating units for each NaPA can be calculated and hence the equivalent PAA Mw can be determined. Also included in Figure 2 are data for citric acid (equiv. n=2.7) and PAA of Mw 2000 (n=27.8). It is obvious that the greatest reduction in τ_{ymax} is obtained with the lowest Mw NaPA at any given additive concentration. Citric acid, being the smallest molecule, exhibits the lowest τ_{ymax}. For example at 0.1dwb% additive concentration, the τ_{ymax} is 250Pa for citric acid, 300Pa for NaPA of Mw 4810 and 380Pa for NaPA of Mw 52000. The τ_{ymax} is 440Pa in the absence of additive. It is also evident in Figure 2 that the extent of τ_{ymax} reduction is much smaller for high Mw NaPA. No reduction in τ_{ymax} was observed for the 57wt% ZrO_2 suspension treated with 0.1dwb% PAA of Mw 750000 (n=10417) as shown in Figure 1(e).

Low Mw additives such as citric acid are more effective in reducing τ_{ymax} because of the absence of bridging flocculation, i.e., a polymer additive is adsorbed onto two or more particles at the same time. Bridging flocculation is only important for high Mw additives and at low additive concentrations[12]. Bridging flocculation is an additional attractive force which enhances the τ_{ymax} of the flocculated ZrO_2 suspensions. The smaller τ_{ymax} reduction exhibited by the high Mw additive in Figure 2 is consistent with bridging flocculation becoming more important for high Mw additives. When the enhancement of τ_{ymax} by bridging flocculation equally balances the reduction by additive coated surface interactions then no reduction in τ_{ymax} will be observed. This may explain the

          Characterization Techniques for the Solid-Solution Interface

Figure 2: Effect of polyacrylate molecular weight on τ_{ymax}.

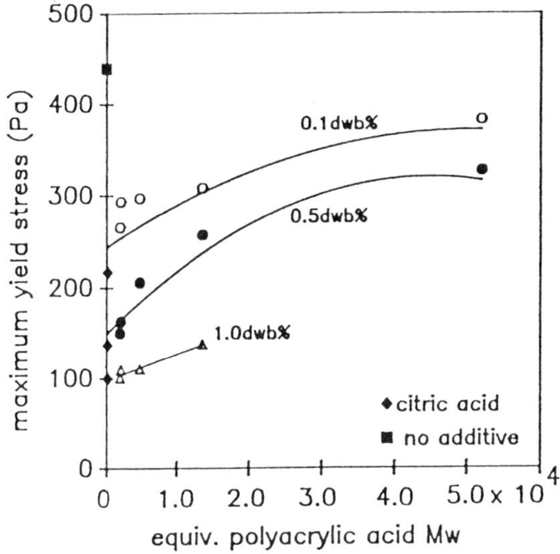

Figure 3: Effect of polyacrylate Mw and concentration on the pH of τ_{ymax}.

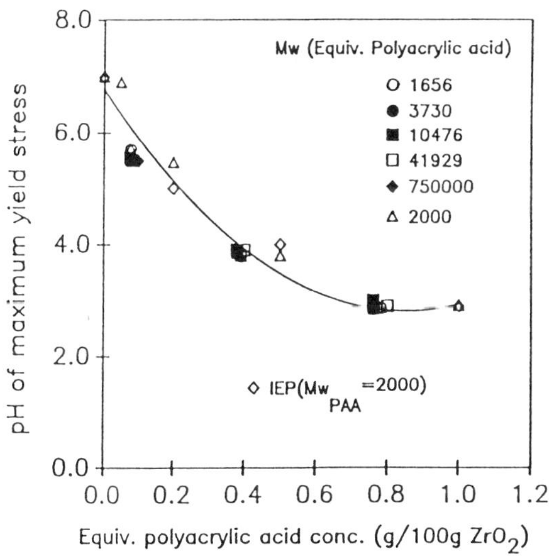

absence of a reduction in τ_{ymax} for the PAA of Mw 750000.

The particle network structure is considered uniform if the flocculated suspension appears smooth or 'homogeneous'. Highly reproducible τ_y-pH behaviour has been obtained with the smoothly flocculated ZrO_2 suspensions [3]. Occasionally a 'grainy' (large aggregates) flocculated suspension was produced which usually occurred at low solids concentration or at high additive concentration of a high Mw polyacrylate. The grains are aggregates of smaller particles. The aggregates may be suspended in a medium of smoothly flocculated suspension. Suspensions containing a high concentration of aggregates generally give a lower yield stress because aggregates may consume a considerable amount of small particles thereby reducing the density of particle-particle bonds. Yielding does not occur within the aggregates and hence the strong particle-particle bonds within the aggregates do not contribute to the yield stress.

The effects of polyacrylate Mw and concentration on the pH of τ_{ymax} are shown in Figure 3. In all cases, the pH of τ_{ymax} decreases with increasing concentration of polyacrylate additives. It appears that polyacrylate Mw does not have a significant effect on the pH of τ_{ymax}. This result suggests that the degree of polyacrylate adsorption at the pH of τ_{ymax} for all Mw is the same or that the adsorption is 100%. The pH of τ_{ymax} of 7.0 without additive is shifted to about 3.0 at 1.0dwb% NaPA for all Mw. Also included in Figure 3 are IEP data for various concentrations of polyacrylic acid of Mw 2000 obtained with the electroacoustic technique. Note that the pH of τ_{ymax} and IEP have the same value for any given polyacrylate concentration.

Figure 4 shows the effect of polyacrylic acid concentration (Mw=2000) on the yield stress-pH behaviour of a 71wt% MEL ZrO_2 (S.A.=3m^2/g) suspension. Like previous results [3], the magnitude and pH of τ_{ymax} are significantly lowered by PAA. The τ_{ymax} is reduced from about 110Pa at 0.0dwb% to about 20Pa at 1.0dwb% PAA, i.e., a 5-fold reduction. The reduction in τ_{ymax} is accompanied by a shift in the pH of τ_{ymax} from a value of 6.5 to about 2.5.

Compared to the τ_{ymax} of 440Pa obtained for Z-Tech ZrO_2 suspension of 57wt% solids, MEL ZrO_2 suspension exhibits a much lower τ_{ymax} of 110Pa despite having a much higher solids concentration of 71wt%. The reason being that MEL ZrO_2 has a much coarser size distribution (D_{so}=0.62μm compared to 0.25μm for Z-Tech powder) compared to Z-Tech ZrO_2. For a given solids volume fraction, the MEL ZrO_2 suspension has less particles per unit area and

     Characterization Techniques for the Solid-Solution Interface

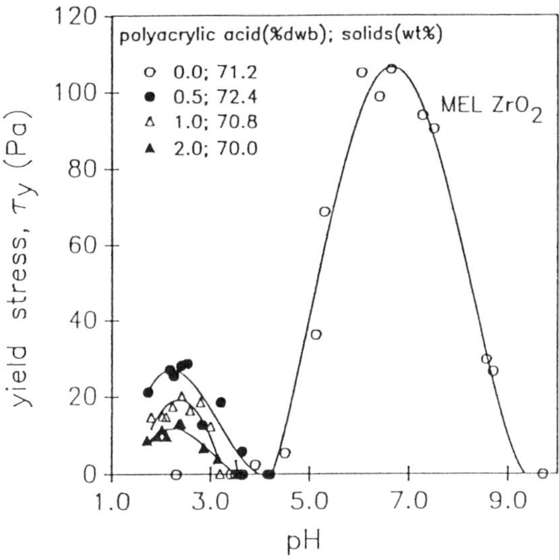

Figure 4: Effect of polyacrylic acid concentration on the yield stress-pH behaviour of 71wt% MEL ZrO_2 suspensions.

Figure 5: Effect of polyacrylic acid concentration on the pH of τ_{ymax} and IEP of both MEL and Z-Tech ZrO_2 suspensions.

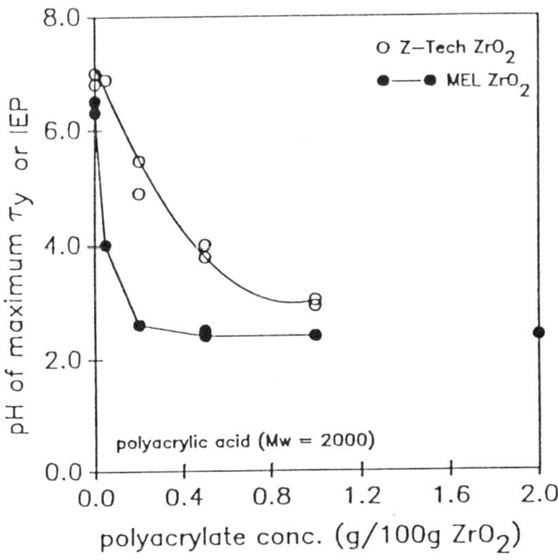

Figure 5: Effect of polyacrylic acid concentration on the pH of τ_{ymax} and IEP of both MEL and Z-Tech ZrO_2 suspensions.

hence a smaller number of particle-particle bonds per unit area. Although the van der Waals attraction between coarser particles is stronger, it is not sufficient to compensate for the decrease in τ_{ymax} caused by a lower number of particle-particle attractions.

Figure 5 shows the decrease in the pH of τ_{ymax} and IEP with increasing concentration of PAA (Mw=2000) for both MEL and Z-Tech ZrO_2. It is very clear that the lower surface area MEL ZrO_2 requires less PAA to reach the limit of IEP or pH of τ_{ymax} shift. The limit occurs at the pH of about 2.5. Presumably, at this pH the ZrO_2 particles are fully covered with a layer of polyacrylate. Figure 6 shows the relationship between the surface coverage of PAA ($gPAA/m^2ZrO_2$) and ΔIEP ($IEP_{no\,PAA}$-IEP_{PAA}) for both MEL and Z-Tech ZrO_2. The result shows that both ZrO_2 exhibited the same relationship between surface coverage and ΔIEP, i.e., the amount of surface coverage of PAA required to achieve a given ΔIEP is the same for both MEL and Z-Tech ZrO_2. Since Z-Tech ZrO_2 is five times larger in surface area, it requires five times more PAA to achieve the same surface coverage. Both Z-Tech and MEL ZrO_2 also show a similar variation in surface charge density with ΔpH (pH-IEP) at a fixed ionic strength of 0.01M KNO_3 in particular at pH below the IEP as shown in Figure 7. These surface charge results may be responsible for both ZrO_2 exhibiting the same PAA surface coverage-ΔIEP relationship.

With the knowledge gained in this and other studies [3,6,7], it should be possible to prepare highly concentrated, low viscosity ZrO_2 suspensions which are so essential in slurry molding applications. Cesarano and Aksay [1] showed that Al_2O_3 ceramics prepared from 0.6 volume fraction (solids) suspensions attained maximum theoretical density at a much lower firing temperature compared to one prepared from dry Al_2O_3 powder. A preliminary attempt using low molecular weight NaPA (Mw=2103) was successful in achieving a low viscosity suspension of volume fraction approaching 0.6. The viscosity-shear rate behaviour of highly concentrated Z-Tech and MEL ZrO_2 suspensions is shown in Figures 8(a) and (b). It can be seen in Figure 8(b) that a maximum volume fraction of 0.57 was achieved for MEL ZrO_2 which has a viscosity of only about 0.5Pas at a shear rate of $100s^{-1}$. The concentration of NaPA used was $0.00033g/m^2ZrO_2$ (0.5dwb% for Z-Tech and 0.1dwb% for MEL) and the resulting suspension pH was about 8.9 for both ZrO_2. The flow beahaviour of the ZrO_2 suspensions was determined using a Bohlin constant stress rheometer.

In summary this study has shown that low Mw adsorbing additives

Figure 6: The relationship between surface coverage of PAA and ΔIEP for both MEL and Z-Tech ZrO_2.

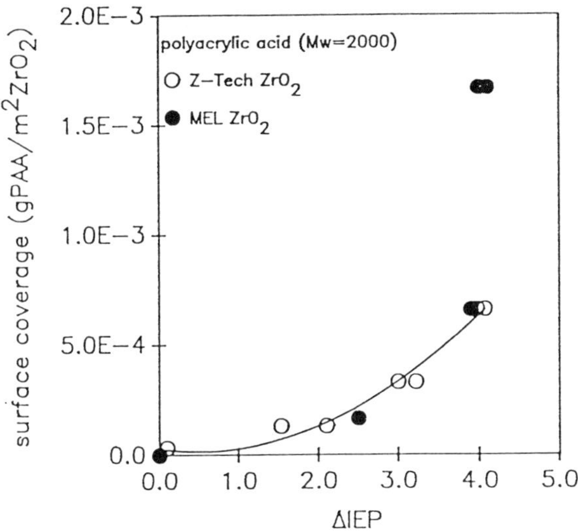

Figure 7: The relationship between surface charge density and ΔpH for both MEL and Z-Tech ZrO_2.

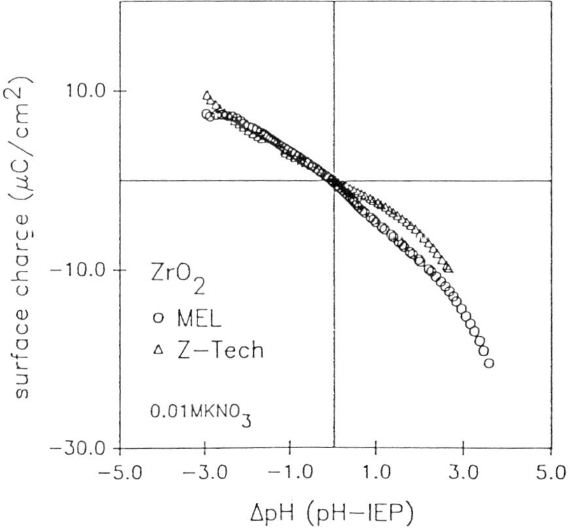

Figure 8a: Effect of solids concentration on the viscosity-shear rate behaviour of Z-Tech ZrO_2 suspensions treated with sodium polyacrylate of Mw 2103.

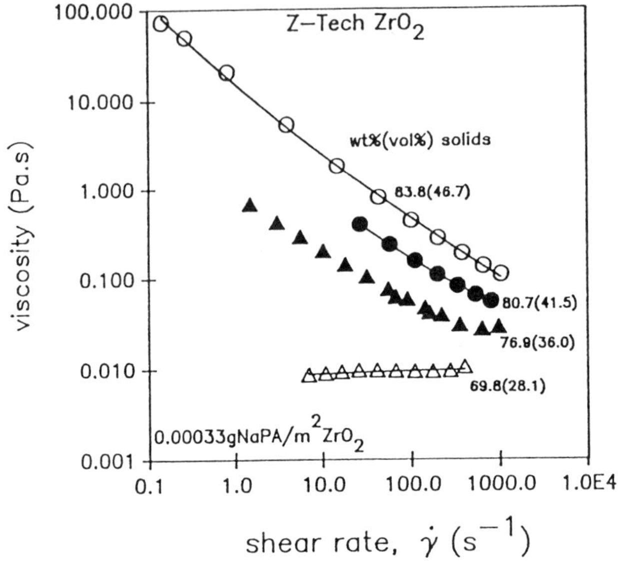

Figure 8b: Effect of solids concentration on the viscosity-shear rate behaviour of MEL ZrO_2 suspensions treated with sodium polyacrylate of Mw 2103.

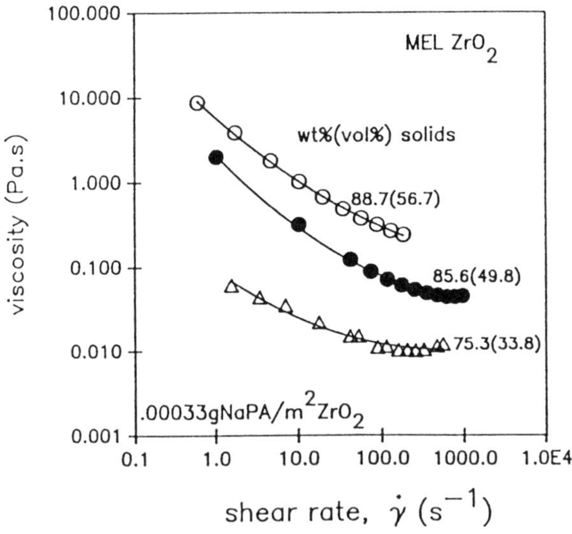

     Characterization Techniques for the Solid-Solution Interface

produced the greatest reduction in τ_{ymax}. The maximum viscosity is also smaller for ZrO_2 treated with low Mw polyacrylate [7]. Contrary to prevailing practice[1,2], it may not be necessary to completely cover the ceramic oxide surfaces with a monolayer of additive in order to achieve better flow properties and processing. Hence, excessive use of additives can be avoided and a combination of additive and pH control may be a more effective method to prepare ZrO_2 suspensions with the desired rheological properties.

Conclusions:

The molecular weight of adsorbing polyacrylate has an important effect on τ_{ymax} of ZrO_2 suspensions. Low Mw polyacrylates are much more effective in reducing the τ_{ymax} of the ZrO_2 suspensions. High Mw polyacrylate is less effective because of significant bridging flocculation. Citric acid, being the smallest molecule, exhibits the greatest reduction in τ_{ymax}.

The extent of IEP or pH of τ_{ymax} shift is determined by the degree of PAA surface coverage for both MEL and Z-Tech ZrO_2. For a given ΔIEP, the surface coverage of PAA required is the same for both ZrO_2.

A low viscosity ZrO_2 suspension of solids volume fraction approaching 0.6 has been achieved using NaPA to modify the surface of the ZrO_2 particles.

Acknowledgement

We would like to acknowledge R. Buscall (ICI Corporate Colloid Science Group) and Allied Colloids for providing and characterising the sodium polyacrylate of different molecular weight, and to ICI Advanced Ceramics for providing the ZrO_2 samples. Our research in the processing of ceramics is funded by the Australia Research Council through a Special Research Centre Grant.

References

1.   J. Cesarano III and I.A. Aksay, ''Processing of Highly Concentrated Aqueous α-Alumina Suspensions Stabilized with Polyelectrolytes'', J. Am. Ceram. Soc., **71** (1988) 1062-1067

2.   E.M. DeLiso, A.S. Rao and W.R. Cannon, ''Electrokinetic Behaviour of Al_2O_3 and ZrO_2 Powders in Dilute and Concentrated Aqueous Dispersions'', Advances in Ceramics, Vol. 21: Ceramic Powder Science,

(1987) 525-535.

3.   Y.K. Leong and D.V. Boger, "Effect of Polyacrylate on the pH and Magnitude of Maximum Yield Stress of m-ZrO_2 Suspensions", Ceramic Transaction, Vol. 19: Advanced Composite Materials, (ed. M.D. Sacks) (1991) 83-91.

4.   A. Bleier and C.G. Westmoreland, "Effects of Adsorption of Polyacrylic Acid on the Stability of α-Al_2O_3, m-ZrO_2 and their Binary Suspension Systems", Process Technology Proceeding 7, Interfacial Phenomena in Biotechnology and Materials Processing, (eds. Y.A. Attia, B.M. Moudgil and S. Chander) Elsevier (1988) 217-236.

5.   M. Hashiba, H. Okamato, Y. Nurishi and K. Hiramatsu, "Dispersion of ZrO_2 in Aqueous Suspensions by Ammonium Polyacrylate", J. Mat. Sci., **24** (1989) 873-876.

6   Y.K. Leong, D.V. Boger and D. Parris, "Surface and Rheological Properties of Zirconia Suspensions", Trans IChemE, **69** Part A, (1991) 381-385.

7   Y.K. Leong, N. Katiforis, D.B.O'C Harding, T.W. Healy and D.V. Boger, "Role of Rheology in Colloidal Processing of ZrO_2", Proceeding Int. Conf. Transport Phenomena in Processing, (Hawaii, March 1992) 10pages - to be published by Technomic Press.

8   Y.K. Leong, M. Prica, P.J. Scales, T.W. Healy and D.V. Boger, "Surface Chemistry and Rheology of ZrO_2 Suspension Containing Polyacrylic Acid", to be published.

9   Q.D. Nguyen and D.V. Boger, "Yield Stress for Concentrated Suspensions", J. Rheol., **27** (1983) 321-349.

10   Q.D. Nguyen and D.V. Boger, "Direct Yield Stress Measurement with Vane Method", J. Rheol., **29** (1985) 335-347.

11   P.J. Scales and E. Jones, ''The Effect of Particle Size Distribution on the Accuracy of Electroacoustic Mobilities'', Langmuir, **8** (1992) in press.

12   R.J. Hunter, Foundation of Colloids Science, Vol.1, Clarendon Press, Oxford, (1987) 489.

Novel and Innovative Characterization Techniques

PORE CHARACTERIZATION OF SiO$_2$ CHROMATOGRAPHIC GELS USING NUCLEAR MAGNETIC RESONANCE

C. Francisco Lorenzano Porras*, P. W. Carr** and A. V. McCormick*
*University of Minnesota, Department of Chemical Engineering and Materials Science
**University of Minnesota, Department of Chemistry.

ABSTRACT

NMR techniques have proven useful for pore structure analysis and related transport properties whereas traditional techniques do not fully account for the topology' of the porespace. To avoid errors due to the topology of the pore space, we have used NMR spin lattice relaxation to calculate the pore size distribution. We have also used the NMR pulsed field gradient to measure the influence of the pore structure on diffusion of the pore fluid and to determine the connectivity of the pores using network models. In combination with traditional techniques, NMR provides much more thorough pore characterization than traditional techniques alone do.

INTRODUCTION

In this work we report the limitations of conventional techniques in characterizing the pore space of a silica chromatographic material. Although useful, none of these techniques is able to measure the effect of dead ends or of tortuosity in slowing the average diffusion coefficient of solutes. Therefore, we use NMR spin lattice relaxation (T_1) and the spin echo NMR self-diffusion experiment to confirm the pore size distribution and to measure the effect of the tortuous pore structure on diffusion of the pore fluid.

Since the rate of spin lattice relaxation for a fluid in the vicinity of a surface is faster than that for the fluid far removed from the surface, NMR spin lattice relaxation measurements of pore fluids have been used to obtain pore size distributions.[1-4] The hydraulic radius distribution thus obtained has shown convincing agreement with conventional techniques but with no error due to pore topology.[5-8]

We also use NMR self-diffusion measurements to find the diffusion structural factor σ, which is defined as the ratio of the diffusivity in the porous medium to the bulk liquid diffusivity. Network models have proved very effective for modeling transport phenomena in porous media by capturing the essential topological features of the pore space such as tortuous paths, dead ends, branching, etc.[9-11] A typical network model is characterized by its coordination number (Z), the number of throats meeting at a pore body.[11,12] Moreover, an irregular network has transport properties nearly identical with a regular network, as long as they have the same coordination number and pore size distribution.[13] Therefore, we quantify the "connectedness" of the pores by estimating the average coordination number exhibited by the diffusivity of the pore fluid.

EXPERIMENTAL

Electron micrographs using low and high magnification were taken over several areas. SEM shows the outer surface and fractured inner surface of the material with resolution up to ~ 100 nm, while TEM allows resolution on the order of a nanometer. Analysis of this information leads to a preliminary model of the pore structure.

Nitrogen adsorption measurements were performed according to recent recommendations.[14-17] The pore size distribution (PSD) is calculated with the method proposed by Barrett, Joyner and Halenda (BJH- method), which treats the pores as cylindrical and neglects errors introduced by pore blockage.[17,18] NMR T_1 relaxation measurements provides an alternative to measure the hydraulic radius for mesopores (radii larger than 5 nm) with *no* pore shape assumption and with *no* error due to pore blockage or constrictions.

 Characterization Techniques for the Solid-Solution Interface

The porous material must be saturated with a fluid possessing a strong NMR signal. Since the application will involve dilute aqueous solutions , we have used water. We employed outgassed deionized water to avoid effects of oxygen and paramagnetic ions, which would interfere with T_1 measurements. The inversion recovery method is used on a Varian 500 NMR spectrometer at a proton frequency of 500 MHz and at room temperature.[19] Delay times ranged from 0.009 to 9 seconds. The relaxation of the protons of water molecules is related to the pore size distribution in the following way

$$M(t) = M_0 \int_{T_{1min}}^{T_{1max}} [1 - 2\, e^{-(t/T_1)}]\, f(T_1)\, dT_1$$

where $f(T_1)$ describes the distribution of spin lattice relaxation times, and, thus, of pore size [1-3,7].

Spin echo pulse field gradient NMR has been applied to study the effect of the pore structure on the rate of diffusion. We wish to detect translational motion on a scale smaller than the particle size. Measurement of intraparticle diffusion is possible if, under the experimental conditions, the root mean square displacement is smaller than the particle dimensions but is large enough to represent isotropic, pseudo homogeneous diffusion.[20,21]

To overcome the limitation of the SEPFG experiment for liquids with low T_2 values we use the stimulated spin echo technique.[20] NMR self diffusion experiments were performed on a Nicolet NMC 1180 NMR spectrometer at a Larmor frequency of 300 MHz. The saturated samples were sealed in a 5 mm tube to avoid evaporation of the water. The field gradient was applied for 1 to 5 ms. The diffusion time Δ was constant during each determination with values between 14 and 504 ms.

DISCUSSION OF RESULTS

SEM and TEM

The low resolution SEM micrograph of figure 1 shows a fine texture and a particle size larger than 70 μm. Figure 2 reveals a very fine microstructure on the scale of the TEM analysis, but it is evident that the SEM and TEM analysis alone do not provide adequate characterization.

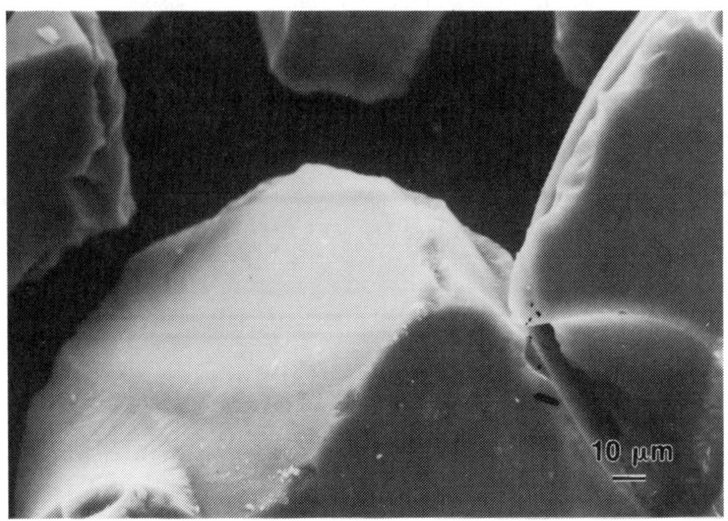

Figure 1 Scanning electron micrograph of the silica gel studied

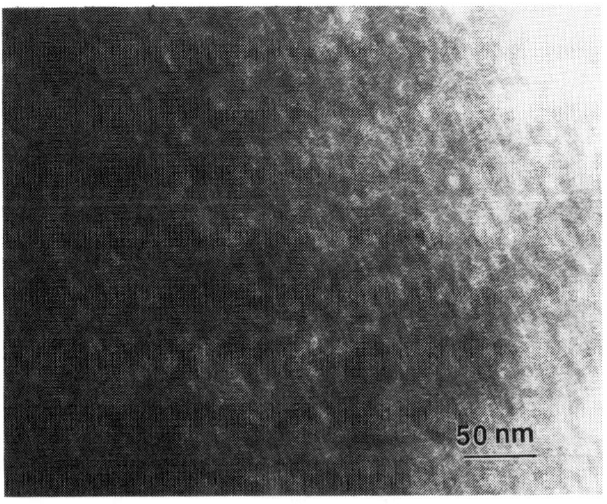

Figure 2 TEM micrograph of the silica gel studied

   Characterization Techniques for the Solid-Solution Interface

N$_2$ adsorption

Figure 3 shows the nitrogen adsorption isotherm for silica gel; it exhibits a type IV isotherm with H$_1$ shape hysteresis, which is associated with a narrow pore size distribution. The isotherm shows low adsorption at low relative pressure (monolayer formation), a gradual increase in adsorption at intermediate pressure (multilayer formation) and, at $p/p_0 > 0.7$, capillary condensation. The hysteresis implies the presence of mesopores (2-10 nm). No evidence of micropores (< 2 nm) is observed in this curve. The surface area is 320 m^2/g and the total pore volume is 60%.

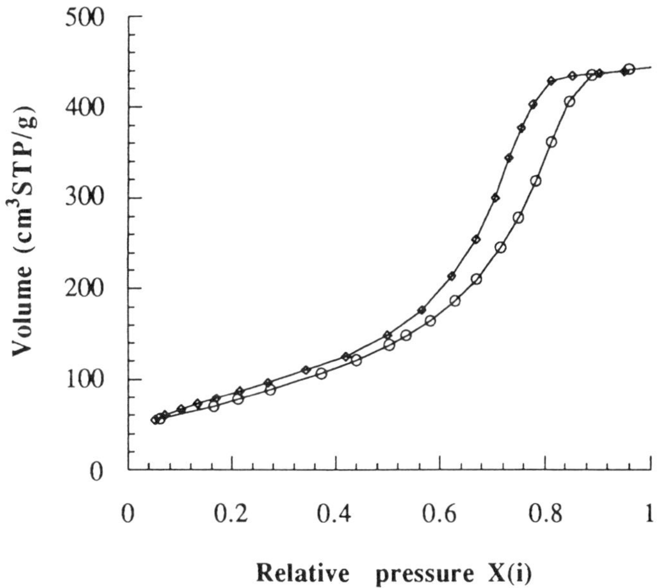

Figure 3 Adsorption-Desorption isotherm for silica gel

Figure 4 shows the narrow pore size distribution calculated from the desorption branch of the isotherm, but more than 70% of the pore volume corresponds to pores with radius between 2 and 4 nm. However, this analysis is subject to uncertainty due to the possibility of constrictions or network effects. It should be noted that for chromatography applications it is important to know whether it is true that a substantial amount of pore volume is only 20 Å in radius, or whether 20 Å constrictions block access to 40 Å pores.

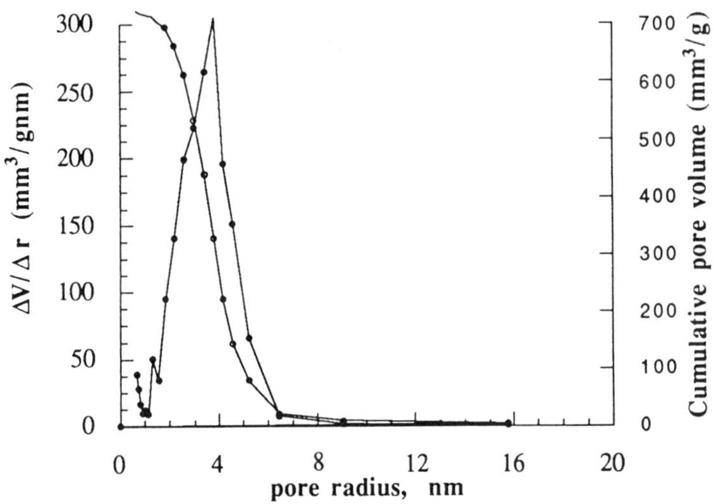

Figure 4 Pore size distribution for silica gel for N_2 desorption .

NMR characterization

Spin lattice relaxation times and the effective pore radius

The relaxation time for silica gel fits a single exponential decay with $T_1 = 2.37 s$. Using values from the literature and extrapolating the relaxation data to 500 MHz using ordinary NMR relaxation theory, the hydraulic radius (2V/SA) comes to 5.5 nm.[7,22] This is somewhat higher than 4.25 nm from adsorption measurements, but the latter may not be accurate due to network effects, which will shift the distribution to smaller pore sizes. We conclude, then, that the characterized silica gel is formed by large pores accessed through smaller constrictions.

     Characterization Techniques for the Solid-Solution Interface

Furthermore, since the signal decay is well described by a *single* exponential decay, we conclude that a diffusing molecule experiences virtually all pore radii in the time scale of the experiment (msec) so the pore undulates in diameter along its axis.

Self-diffusion coefficient

The diffusivity for water in silica gel is shown in figure 5.   The relative diffusivity is almost constant for diffusion times of 100-500 ms, but at shorter diffusion times the relative diffusivity seems to increase. Examination of the root-mean-square displacements, though, assures us that intraparticle diffusion is dominant for all these measurements since the water molecules are still moving over distances much smaller than the particle size.   At very small diffusion times the water molecules are probably not exploring a sufficiently large portion of the particle to give a statistical representation.

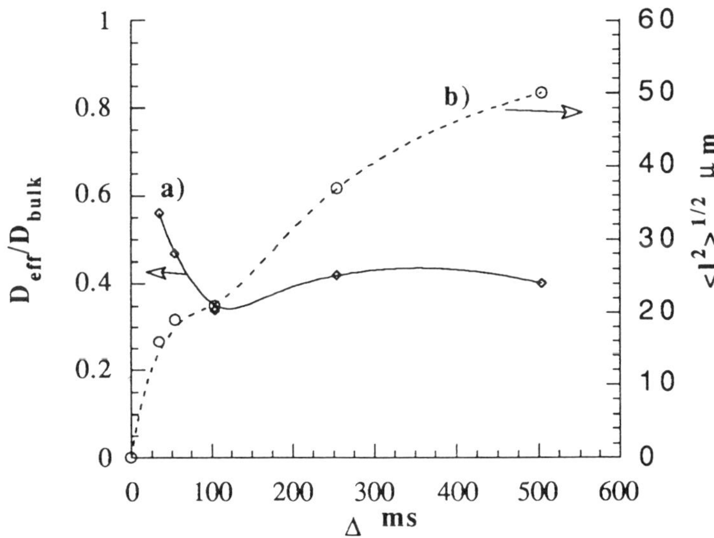

Figure 5 a) Relative self-diffusion coefficient of water in the silica gel as a function of the observation time Δ; b) Dependence of the root mean square displacement on Δ.

Because the *rms* displacement is smaller than the particle size, the intraparticle structural factor can be taken as 0.4 ε, where ε is the porosity. From the solution to the ordinary diffusion equation in a Bethe network given by Reyes, we infer that the coordination number, Z, is less than 4, which implies a very poorly connected pore structure.[10]

CONCLUSIONS

We have seen the limitations of conventional techniques to measure the effect of the topology of the pore structure (branching, dead ends, accessible surface area, blocked pores, tortuosity, etc) in slowing the effective diffusivity of solute molecules. Spin echo self diffusion NMR is a practical and effective method to do such evaluation by determining the average coordination number of the porespace using network models. Only with a combination of techniques it is possible to characterize the porespace sufficiently well for applications such as chromatography.

[1] J. A. Brown, L. F. Brown, J. A. Jackson, J. V. Milewski and B. J. Travis SPE/DOE 10813 *Unconventional Gas Recovery Symposium,* May 16-18, 1982

[2] R. J. S. Brown and I. Fatt, 1956, *Petroleum Trans. AIME,* Vol. 207, p. 262 .

[3] J. A. Brown, L. F. Brown, J. A. Jackson, 1981, *Low Permeability Symposium,* D. Colorado, May 27- 29.

[4] J. A. Glasel and K. H. Lee, 1974, *Journal of the America Chemical Society,*Vol. 96, No.4, p. 970-978.

[5] The hydraulic radius is defined as r_h = surface area/volume.

[6] D. P. Gallegos, K. Munn, D. M. Smith and D. L. Stermer, 1987, *Journal of Colloid and Interface Science,* Vol. 119, No. 1, p. 127-140.

[7] D.P. Gallegos, *Application of NMR Spin Lattice Relaxation Measurements to the Analysis of Pore Structure,* M. Sc., University of New Mexico, 1987.

[8] D. P. Gallegos, D. M. Smith and C. J. Brinker, *J. of Colloid and Surface Sci.,* Vol. 124, No. 1, p. 186-198.

[9] K. K., Mohanty, J.M. Ottino and H.T.Davis,1982,*Chemical Engineering Science,* vol. 37, No. 6, p. 905-924.

[10] S. Reyes , *Ph. D.* Thesis University of Minnesota, 1985.

[11] M. Sahimi , G. R. Gavalas and T.T. Tsotsis, *Chemical Engineering Science,* 1990, vol. 45, No. 6, p. 1443-1502

[12] J. Koplick , 1982, *Journal of Fluid Mech.,* Vol. 119, p. 219-247

[13] G. R. Jerauld , *Ph. D. Thesis,* University of Minnesota , Minneapolis 1985

[14] ASTMD4222- 83 *Standard test method for determination of Nitrogen Adsorption and Desorption Isotherms of Catalysts by Static Volumetric Measurements,* May 1983.

[15] ASTMD3663- 84 *Standard Test Method for Surface Area Analysis,* June 1984.

     Characterization Techniques for the Solid-Solution Interface

[16] IUPAC *Reporting Physisorption data for gas - solid systems* (recommendations 1984). Pure and Applied Chem. Vol. 57, No.. 4, 1985.

[17] ASTM D 4641-88 *Standard Practice for Calculation of Pore Size Distribution From Nitrogen Desorption Isotherms*, January 1989.

[18] E.P Barret, L.G Joyner and P.P. Halenda, 1951, *J. American Chemical Society*, Vol. 73, p. 373-380.

[19] E. Fukushima and S. B. W. Roeder, *Experimental Pulse NMR*, Addison-Wesley Pub. Co., Massachusetts, 1981.

[20] J. E. Tanner, 1978, *The Journal of Chemical Physics*, Vol. 69, No. 4, p. 1748-1754.

[21] P. T. Callaghan, 1984, *Aust. Jurnal of Physics*, Vol. 37, p. 359-387.

[22] C. L. Glaves, G. C. Frye, D. M. Smith, C. J. Brinker, A. Datye, A. J. Ricco and S. J. Martin, 1989 *American Chemical Society, Langmuir*, Vol. 5, No. 2, p. 459-466.

TEMPERATURE PROGRAMMED DESORPTION AS A SURFACE CHARACTERIZATION METHOD OF CERAMIC POWDERS

M. Kawamoto, K. Ishizaki and C. Ishizaki
Nagaoka University of Technology, Nagaoka, Niigata 940-21, Japan

ABSTRACT

The importance of surface characterization of raw ceramic powders is well known. Temperature programmed desorption (TPD) is a powerful method to analyze sophisticated ceramic powder surfaces. The present paper explains the details of the equipment and analytical procedure to obtain TPD spectra of ceramic powders. The advantages and disadvantages of the present equipment are also discussed.

INTRODUCTION

Ceramics unlike metals are mainly produced by sintering, which implies that the surface structure of the raw powder is a controlling factor to determine the final characteristics of sintered ceramics. Many methods of surface characterization have been proposed and applied in the ceramics field.

In the study of ceramics, a knowledge of the surface chemical composition is not enough to understand the different behaviors encountered. Furthermore, it is important to obtain information not only of a thin surface layer, but also about deeper layers. This information is of relevant importance, for instance, for ceramics which have grain boundary structures and therefore their properties may be governed by the deeper surface layer of the raw ceramic powders.

To overcome the chemical surface modifications that may be introduced in characterizing solid surfaces through solid-solution interphase techniques, the authors implemented a TPD method for the study of ceramic raw powders which at the same time provides information about the compounds bonded to the surface, their chemical bonds, and the deeper surface structure by desorption of the surface layer.

THE TPD TECHNIQUE

The TPD technique was described first in 1963 in a paper by Amenomiya and Cvetanovic (1). Adsorbed gases were desorbed in a programmed manner from a conventional catalytic material into a stream of an inert carrier gas and their concentration in the carrier gas was continuously monitored by a sensitive gas chromatographic detector such as a hot filament or a thermistor detector. Thermal desorption of gases from solids can be used as a simple technique to determine bonds between gases and solids. The flash filament method, a specific version of thermal desorption involving the use of ultra-high vacuum to follow desorption of an adsorbate from a rapidly heated filament, is a well established technique that has provided a great deal of valuable information on adsorption of gases on clean metal surfaces (2). The basic principle is the same as in the flash filaments and can indeed be any type of solid catalytic material.

It is possible to obtain not only information dealing specifically with the desorption process but sometimes also information on the catalytic reactions on the catalyst surface. At high temperatures, surface groups decompose, therefore the spectra also give information about the possible surface functional groups. Since 1963, an increasing number of papers based in part or entirely on the application of this technique have been published (2).

A record of the concentration of the desorbed materials as a function of time in the course of temperature programmed desorption may be referred to as a TPD spectrum. The spectrum generally consists of one or more peaks. The shape of the peaks and the position of the peak maxima on a temperature scale are related in a fundamental way to the desorption process and therefore provide information on the manner in which the gas is adsorbed on the surface species. Extraction of this information in an explicit form depends on a valid interpretation of the spectra.

The basic equations describing thermal desorption from an energetically homogeneous surface were given first in 1958 by Smith and Aranoff (3), and have been subsequently discussed and extended by several other workers such as Ehrlich (4), Redhead (5), Carter (6), Grant and Carter (7), Cvetanovic and Amenomiya (8), Yakerson, Rozanov, and Rubinshtein (9). A detailed discussion of the theoretical background of the TPD technique has also been given by Cvetanovic and Amenomiya (8).

There are mainly two kinds of TPD equipment, gas chromatograph (GC)-type and vacuum-type (10). The vacuum type detects desorbed molecules under ultra high vacuum (UHV) or high vacuum (HV) states by mass spectroscopy. The GC type analyzes desorbed gases by GC in an inert gas, e.g., He. The detailed descriptions appear elsewhere (5-7, 11-15).

Figure 1 (a) shows a schematic representation of vacuum-type TPD equipment (16). It consists of a gas inlet for adsorption, an adsorption cell (C), a vacuum gage (P), a gas analyzer (Q), and outlets to vacuum pumps (I_1-I_3). A background vacuum value much lower than desorption pressure of gases should be achieved. Usually the pressure obtained by pumping of ion or diffusion pumps is

     Characterization Techniques for the Solid-Solution Interface

less than 10^{-6} Pa. High conductance and bakeable (at 200°C-250°C) design improves the resolution of the equipment. The sample powder is located in a bucket at the center of the cell for homogeneous heating. For a precise temperature measurement, the thermocouple should contact the sample. Usually the sample temperature is increased linearly from outside of the vacuum chamber by a temperature control system. To increase the conductance and resolution, a large diameter vacuum chamber must be used. For the mass detection, usually a mass spectrometer is attached.

Figure 1 (b) shows the equipment of a GC-type TPD system. It consists of an inert gas inlet in a reactor chamber (A), flow adjuster capillary tubes (C_1, C_2), a desorption and an adsorption cell (C), and a thermal conductivity cell detector (D). Impurities in the carrier gas and leaking from air must be avoided. Basically this TPD system is similar to normal GC equipment.

Generally vacuum type TPD equipment is suitable for quantitative analysis, e.g., to obtain order of desorption mechanism and activation energy of desorption. The GC-type is used to identify desorbed species qualitatively, e.g., catalytic reactions.

For the surface analysis of ceramic raw powders, the vacuum-type equipment was chosen because quantitative analyses and evaluation of surface species themselves are possible. However the normal equipment shown in Figure 1 (a) is not convenient for sample changes. In addition, the heating efficiency and temperature control from the outside furnace are not satisfactory and the obtained spectra include gases from the glass cell. For this reason, a special TPD equipment was designed and constructed by the authors.

TPD EQUIPMENT

Vacuum Chamber

Figure 2 shows schematically the vacuum chamber of the TPD equipment designed and constructed by the authors. This assembly consists of two vacuum chambers, three vacuum pumps, a quadruple mass spectrometer, and a sample transfer system. For a TPD experiment, a powder sample in a cup (mentioned later) is placed into the first chamber located above the gate valve 1 (GV1). This chamber is pumped from air to 5×10^{-6} Pa by a turbo molecular pump (Model ST-160CA, Osaka Vacuum Co. Ltd.) with an oil rotary pump. This chamber also can be used for adsorption experiments. The sample is transferred into a furnace (discussed in the next section) in the second or main chamber by a transfer system. The main chamber is pumped by an ion pump (Model PS-1000, Thermionics Laboratory, Inc.) and a titanium sublimation pump (Model PS-500, ibid). The pressure achieved is 2×10^{-8} Pa, which supports high resolution of this TPD system. To obtain high conductance in the main chamber, the diameter is relatively large, about 300 mm (the flange type is ICF117). This design significantly decreased the setting up time of the sample.

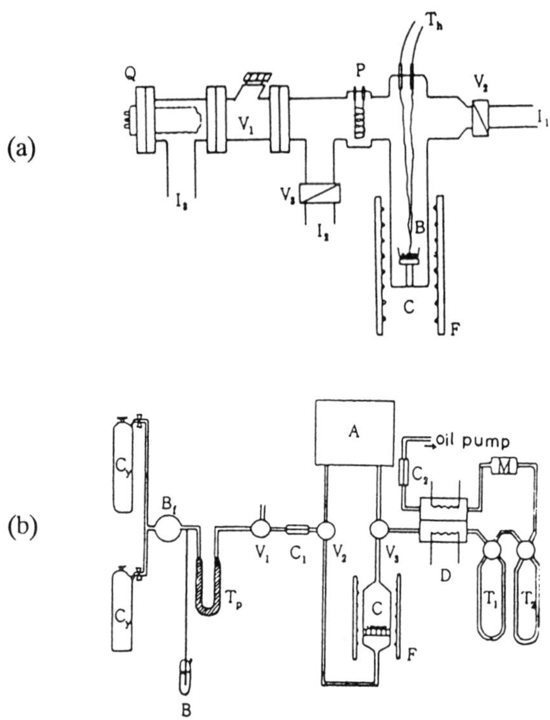

Figure 1. General TPD equipments (16). (a) vacuum type TPD equipment, Q: quadruple mass spectrometer, V_1-V_3: valves, P: vacuum gauge, B: bucket, T_h: thermocouple, F: furnace, I_1-I_3: to vacuum pumps, and C: adsorption and desorption cell, and (b) GC-type TPD equipment, A: reactor, gas inlet and vacuum valves, V_1-V_3: valves, D: thermal conductivity cell, C_1 and C_2: capillary tubes, M: Manometer, T_1, T_2, and T_p: traps, C_y: carrier gas cylinder, B: bubbler, F: furnace, C: desorption and adsorption cell, and B: buffer.

     Characterization Techniques for the Solid-Solution Interface

Furnace

The TPD furnace designed and made by the authors is shown in Figure 3 (a). It consists of six co-axial cylinders of reflectors (inner diameter: 10 mm) made of stainless steel (0.1 mm thickness), a tantalum spiral heater (diameter: 0.3 mm), a stainless steel sample table, and two pairs of thermocouples. The power of the heater is supplied by a proportional integral differential (PID) controller with a temperature regulator (Tosoku Research Co. Ltd.). The maximum temperature in vacuum is 1200 K, which is the degassing temperature. The pressure at this temperature is 10^{-6} Pa, because gases desorb from the stainless steel reflectors, the furnace was kept at 1200 K for 1 hour before TPD experiments. TPD experiments are conducted at temperatures up to 1000 K.

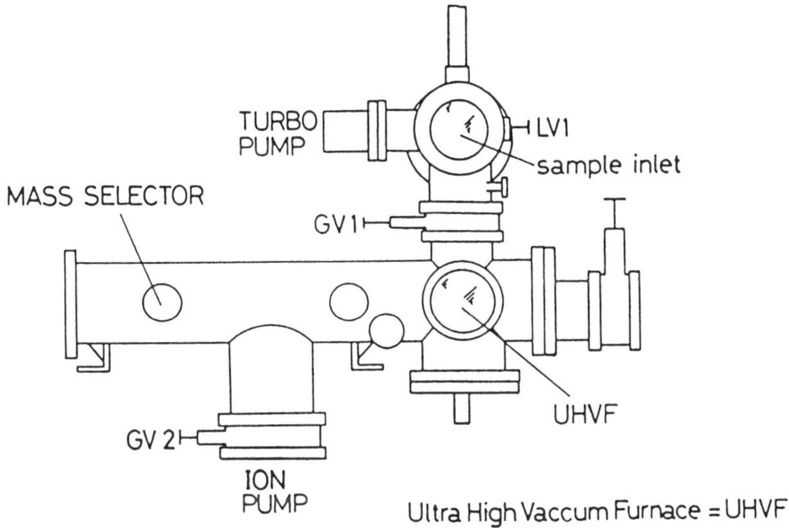

Figure 2. TPD UHV chamber.

A vacuum of 10^{-8} Pa is achieved. The first vacuum chamber of 10^{-6} Pa makes it possible to exchange samples in short time (~ two hours). Since the furnace is located in the main chamber, the heating efficiency and temperature controlling ability are improved.

The significant difference between the two furnaces shown in Figures 1 and 3 is the location. In the case of the former, the size is rather large with indirect heating, then there is delay of thermal conduction. On the contrary, in the case of the present furnace, the size could be decreased with close direct heating, and the delay of heat is smaller than the former one.

Figure 3 (b) shows the cylindrical powder container made of stainless steel. The diameter is 2.5 mm and the wall thickness is 0.08 mm. This container is transferred on top of the sample table, which has direct contact to the sample thermocouple as in Figure 3 (a). In other words, the sample thermocouple contacts directly to the center of the bottom surface of this powder container. During the TPD experiments desorption of gases from this container is negligible, because of small surface area in comparison with that of the sample powder. This was checked by a TPD experiment of the container without sample.

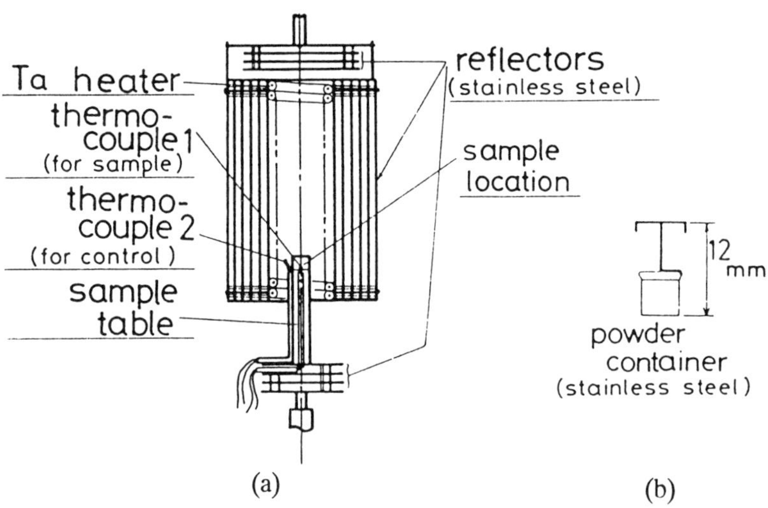

Figure 3. (a) TPD furnace and (b) powder container.

Six coaxial stainless steel reflectors guarantee good insulation of heat. The tantalum heater can resist rapid heating. Two sets of thermocouple are located at the most suitable places for temperature controlling, and measurement.

Control System

The temperature of the furnace and the data of desorbed gases are controlled by a computer. Figure 4 indicates the data flow diagram of this system. This control system consists of the furnace, the mass spectrometer, the temperature controller, a personal computer (PC9801E, NEC Co.), and an X-Y recorder. The dotted box indicates the main chamber shown in Figure 2. The temperature of the furnace is controlled by the temperature controller using the thermocouple 2 in Figure 3 (a). The sample temperature is detected by the thermocouple 1 in Figure 3 (a) and input into the computer after amplification. The sample and the controller temperatures are plotted on the X-Y recorder. The desorbed gases are detected by the mass spectrometer. The computer gives the mass number to analyze, and receives the partial pressures of the gases as input data. All the data obtained are processed by the computer.

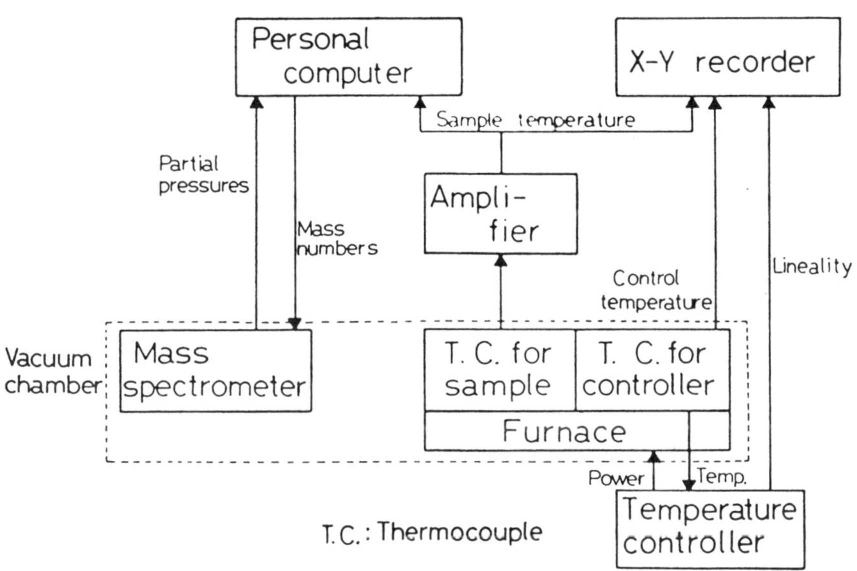

Figure 4. Schematic diagram of the control system. The mass spectroscope probe and the furnace are located in the vacuum chamber. The voltages obtained from them are processed by the computer. Many desorbed gases can be detected and their data are processed simultaneously.

Advantages and Disadvantages of the System

The advantages of this system are: (1) the sample change and transfer are very easy and quick, (2) small furnace is located in the chamber (inner furnace), the heating is fast and direct, and the temperature control is precise, (3) the resolution is very high due to the high vacuum conductance chamber (large size of diameter) and the very low pressure (10^{-8} Pa), (4) because all data are controlled by a computer, many molecules can be analyzed simultaneously and quickly, and (5) adsorption experiments are also possible in the first chamber.

On the contrary there are a few disadvantages as follows: (1) the sample amount is relatively small because the furnace is small, (2) the upper limit of temperature use is 1000 K because the furnace is made of stainless steel and heater of tantalum, and (3) weak bonded molecules can not be detected due to the high vacuum.

TPD ANALYSIS

Procedure

The furnace located in the UHV system is kept at 1150 K for 30 minutes in the UHV system for degassing before TPD experiments. The handling of the powders is always performed in a clean room ($23\pm2°$C and $52\pm3\%$ humidity). For the TPD experiments, a fixed amount (17.6 mg) of sample powder is weighed in a stainless steel cup (diameter: 2.5 mm, height: 4 mm, thickness: 80 μm) and placed in the first vacuum chamber in Figure 2. After 1 hour vacuum (pressure lower than 5×10^{-6} Pa) the sample is transferred into the furnace in the main chamber. The TPD experiment is performed after keeping the sample in the chamber for 2 hours (pressure $\sim 2 \times 10^{-8}$ Pa). The temperature of the furnace is increased linearly (50 K/min) by the TPD temperature controller (Figure 2). During the linear temperature increase, desorbed gases are detected by the quadruple mass spectrometer (Figure 2). The temperature and partial pressure of the desorbed gases are input into the computer memory at the same time.

Calculation of the Number of Desorbed Molecules

By the TPD experiments, desorption behavior of molecules called TPD spectra is obtained. From these spectra the numbers of desorbed molecules (N) can be calculated by using the following equation,

$$N = \frac{S}{A*kT} \int_{t_x}^{t_y} P\,dt$$

   Characterization Techniques for the Solid-Solution Interface

where A is the sample surface area, k the Boltzmann constant, P the partial pressure of the molecules in the system by desorption, S the pumping speed, T the temperature in the system, and times t_x and t_y show the range of integration (10).

The described equipment and analytical procedure was used for the characterization of raw and modified silicon nitride powders. The results have been published elsewhere (16-18).

REFERENCES

1.   Y. Amenomiya and R.J. Cvetanovic, "Application of Flash-Desorption Method to Catalyst Studies I Ethylene-Alumina System," *J. Phys. Chem.,* **67**, 144 (1963).

2.   R.J. Cvetanovic and Y. Amenomiya, "A Temperature Programmed Desorption Technique for Investigation of Practical Catalysts," *Catalysis Reviews,* **6**, 21 (1972).

3.   A.W. Smith and S. Aranoff, "Thermodesorption of Gases from Solids," *J. Phys. Chem.,* **62**, 684 (1958).

4.   G. Ehrlich, "Modern Methods in Surface Kinetics Flash Desorption, Field Emission Microscopy and Ultrahigh Vacuum Techniques," *Advan. Catal.,* **14**, 256 (1963).

5.   P.A. Redhead, *Vacuum,* **12**, 203 (1962).

6.   G. Carter, *Vacuum,* **12**, 245 (1962).

7.   W.A. Grant and G. Carter, "Thermal Desorption of Attached Gas from Surface Sites Processing a Uniform Distribution of Activation Energies," *Vacuum,* **15**, 13 (1965).

8.   R.J. Cvetanovic and Y. Amenomiya, "Hydrogenation of Olefin on Alumina I Active Sites for Hydrogenation of Ethylene," *Advan. Catal.,* **17**, 103 (1967).

9.   V.I. Yakerson, V.V. Rozanov, and A.M. Rubinshtein, *Surface Sci.,* **12**, 221 (1968).

10.   Y. Inoue, *Shokubai (Catalysis),* **24**, 225 (1982).

11.   I. Yasumori, *Kagaku to Kogyo (Chemistry and Industry),* **19**, 1208 (1966).

12.   Y. Amenomiya, "Shokubai Jikken Manyuaru," (Manual of Catalysis Experiments), p. 15, Edited by Shokubai Gakkai, Maki Shoten, Tokyo, 1971.

13.   M. Smutek, S. Cerny, and F. Buzek, *Advan. Catal.,* **24**, 343 (1975).

14.   S. Tsuchiya, *Shyokubai (Catalysis),* **21**, 79 (1979).

15.   I. Toyoshima and I. Yasumori, "Shinjikken Kagaku Kouza 18, Kaimen to Koroido," (New Experimental Chemistry, Vol.18, Interface and Colloid), p. 136, Maruzen, Tokyo, 1979.

16.   M. Kawamoto, K. Ishizaki, and C. Ishizaki, "Characterization of Silicon Nitride Powders by Temperature Programmed Desorption," in "Euro-

Ceramics, Vol.2, Properties of Ceramics", pp. 120-124, Edited by G. de With, R.A. Terpstera, and R. Metselaar, Elsevier Applied Science, London and New York, 1989.

17. M. Kawamoto, C. Ishizaki, and K. Ishizaki, "Fluidity-Increasing Behaviour of Silicon Nitride Powder by Aqueous Washing", *J. Mater. Sci. Lett.*, **10**, 279 (1991).

18. K. Ishizaki, M. Kawamoto, and C. Ishizaki, "Improving Flow Behavior of Silicon Nitride Powder by Surface Modification," in "Surface Modification Technology IV," ed. by T.S. Sudarshan, D.G. Bhat, M. Jeandin, and J.F. Braza, pp. 655-664, The Minerals, Metals and Materials Society, Warrendale, PA, 1991.

DETERMINATION OF MOISTURE CONTENT IN FINE CERAMIC POWDERS USING THE KARL-FISCHER (AQUAMETRY) METHOD

S. Venigalla, J.F. Opalko, and J.H. Adair,
Department of Materials Science and Engineering,
University of Florida, Gainesville, FL 32611.

ABSTRACT

The determination of water in ceramic powders by the Karl Fischer titration method is presented. Karl Fischer reagent (KFR), consisting of a mixture of pyridine, iodine, sulphur dioxide, and methanol, reacts stoichiometrically with water. A given volume of KFR will react with an exact volume of water. In an excess of KFR, the electrical resistance of the solution drops to less than a thousand ohms, facilitating the exact determination of moisture content in a given amount of sample using dead-stop technique in an aquametry apparatus. Standard analytical procedures have been developed to determine the moisture content in fine powders of advanced ceramic materials such as SiC and B_4C using Karl Fischer method. This method can be effectively employed as a quick, reliable, and low-cost tool to determine moisture content during ceramic powder characterization.

INTRODUCTION

The use of Karl Fischer reagent to determine the water content in organic and inorganic substances has been well exploited[1-5]. The aquametry apparatus[1] consists of a glass reaction vessel with a three-port adaptor, a self-zeroing buret, a filling device, a jar to hold the K-F reagent, and drying tubes. In addition to this base unit, it has a magnetic stirrer and a meter-electrode. The Meter-Electrode is a self powered conductance meter, sensitive to three micrograms of water, and has a plug-in platinum electrode. The face of the meter has a yellow area to indicate the

---

1 Aquametry II Apparatus, Labindustries, Inc., Berkeley, CA.

---

water end-point and a brown area to indicate the KFR end-point. The transfer of K-F reagent from stock bottle to burette is made through a teflon connector with dry air under pressure generated by squeezing a rubber bulb. A liquid sample may be introduced into the reaction vessel through the sample port by removing the penny head stopper. A powder sample may be introduced into the vessel from the adapter, exposing the conveniently large opening of the vessel. However, the vessel should be screwed back into the adapter very quickly to avoid the entrance of the moisture into the apparatus. Also, it is not a very accurate way of transferring small quantities of fine powders. Alternatively, the powder samples may also be added by either dissolving or suspending in a solvent. The later method has been employed in the present work.

The dead-stop aquametry technique involves three main steps. Firstly, the selected solvent (approximately 50 ml) is added to the reaction vessel through the sample port. Then the solvent (methanol is the most widely used solvent in aquametry) is titrated with suitably diluted K-F reagent to remove the water (when the solution just changes its color from yellow to dark brown. This is known as the K-F end-point. At this point one has to note and record the indication on the Aquametry Meter-Electrode conductance meter.

Titration can be terminated at any point in the brown K-F zone, provided that it is always terminated at that same end point. Solution should be allowed to stabilize at the end point for at least one minute. If the meter indication drops back to the excess water side of the end-point, the sample should be left in the reaction vessel for a longer period before titrating. A shifting end-point may indicate that the apparatus is not quite dry. Letting the apparatus stand with K-F reagent in the vessel will dry it out, since K-F reagent is a very effective desiccant.

The next step is to standardize the water equivalence of the K-F reagent. Normally this needs to be done only once a day, or when changing the K-F reagent supply. This is performed by adding a known amount of water (usually 10-20 mg) to the vessel and titrate to the end-point as described above. Compute the K-F reagent water equivalence, F = mg of water added/ml of K-F reagent required to return to the same end point. To the reaction vessel, still at the end-point, add a weighed amount of the sample which is of sufficient size to contain 15-25 mg of water. A sufficiency of water is indicated by the solution color change to light yellow. The sample is titrated to the original end-point and the buret reading of the K-F reagent is recorded. The water content of the sample is calculated as follows :

Water (%) = {Volume of K-F reagent (ml) x F x 0.1}/ Weight of the sample(g)

          Characterization Techniques for the Solid-Solution Interface

MATERIALS AND METHODS

The Karl-Fischer reagent[2] is diluted (with dilutant for K-F reagent[1]) to a suitable degree (usually 1:1 or more) depending on the expected degree of moisture content in the sample. This requires prior knowledge of the approximate level of moisture in the given sample and also the weight of sample to be titrated. This is important for obtaining an accurately measurable reading for the amount of K-F reagent added to neutralize the moisture content in the sample during the titration. For example, if the K-F reagent is too strong, and the moisture in the sample is too low, then the burette reading for the volume of K-F reagent titrated will be too small to be measurable. Also, a dilute reagent gives enough flexibility for the operator to turn the burette on and off, particularly in the case of manually fed titration apparatus.

Standardization of the K-F reagent is performed by adding a known amount of water (typically 5 mg of H_2O) in the form of standard water (1mg of H_2O / ml)[1] to the pre-titrated solvent (methanol[1]). Titration of this mixture gives the amount of K-F reagent required to neutralize a known amount of moisture. For example, 4.13 ml (average of five readings) of K-F reagent neutralized 5 mg of H_2O. Thus, 1 ml of K-F reagent neutralizes 1.21 mg of H_2O.

Due to inherent problems in transferring the small and precise amounts of fine powder samples into the titration vessel, it has been decided to suspend the powders in standardized methanol (typically 0.2 g / 5 ml). The suspensions of 5 ml each are individually prepared in small teflon cups. The use of teflon cups minimizes any residual suspension being left in the container after transferring the sample into the titration vessel. Four such individual suspensions are prepared for each powder sample using high precision measuring devices[3]. Standardization of methanol for this purpose is performed by titrating known volume of dry methanol with K-F reagent. Each ml of dry methanol typically contained 0.28 mg of H_2O.

The suspensions of powder samples prepared as above are added to the titration vessel containing pre-titrated solvent and then titrated with the K-F reagent. The amount of moisture contributed by 5 ml of standardized methanol is deducted from the total moisture reading obtained from each K-F titration to calculate the moisture content in the sample. Titration for each sample is repeated four times with individual suspensions.

---

[2] Fisher Scientific Company, Fair Lawn, NJ. All chemicals are Certified Reagent Grade, unless otherwise specified.

[3] Fisher Scientific A-250 Analytical Balance, Fisher Scientific Company, Fair Lawn, NJ.

Adjustable volume Finnpipette®, 1-5 ml, Cole-Parmer Instrument Company, Chicago, IL.

RESULTS AND DISCUSSION

The as-received methanol contains large amounts of moisture and is not suitable for use as the solvent in K-F titrations since prohibitively large quantities of K-F reagent are required to neutralize the solvent. Also, the accuracy and consistency of the determination of water in powder samples is rather poor when the samples are suspended in such a solvent. For these reasons, the as received methanol is desiccated by soaking with zeolite[4] beads. Dehydrated methanol is then standardized with K-F reagent and used as the solvent.

The results of the K-F titrations performed on commercial grade SiC and B_4C powders are shown in Table 1. The K-F end-point is not quite stable in the third and fourth titrations of each sample, indicating a substantial interference from the previously titrated samples and possibly the moisture, gradually entering from the ambient atmosphere. This instability might have had significant contribution to the increased moisture content values in the last two readings of each sample. It is therefore recommended to change the solvent base in the titration vessel after every other titration. Further experiments are being performed to confirm this effect. The results are, otherwise, fairly consistent and seem to be accurate. Suggested future work include the titrations with standard samples (ceramic powders with known amounts of moisture) and comparison with other methods of moisture determination like TGA.

CONCLUSIONS

1. Aquametry titrations using Karl-Fischer reagent are successfully employed to determine moisture content in fine ceramic powders with reasonable accuracy and consistency. This method has potential applications in ceramic powder characterization as a quick, low cost, and reliable tool to estimate water content.

2. Dehydration of methanol using zeolite molecular sieves improves the stability of the titration end-point and thereby the consistency and accuracy of the moisture determinations.

3. Transferring the small and precise quantities of fine powders to the titration vessel is non-trivial. Individual suspensions of powder samples in standardized solvent are used in these experiments to minimize the error as well as the moisture entering from the ambient atmosphere.

4. Accumulation of previously titrated suspensions seems to interfere with further titrations and affect the stability of the end-point which is evident in the later

---

[4] Molecular Sieves: A Synthetic Silico Aluminate Zeolite, Grade 518, 4Å Effective Pore Size, 10-16 Mesh Beads, Davison Chemical, Baltimore, MD.

Sample No.	Description	Particle Size (μm)	Sample Weight (g)	Moisture (wt %) Raw Data	Mean
1	SiC	0.4	0.214	1.12	1.53
			0.207	1.12	
			0.211	1.72	
			0.210	2.14	
2	SiC	2.1	0.211	1.95	2.40
			0.210	1.72	
			0.202	2.82	
			0.211	3.12	
3	SiC	4.2	0.218	1.29	1.97
			0.203	1.86	
			0.209	2.15	
			0.201	2.57	
4	B_4C	0.9	0.204	1.92	2.51
			0.212	2.13	
			0.202	2.56	
			0.216	3.41	

Table 1. K-F titration data of fine ceramic powders (SiC and B_4C)

part of titrations for each sample. It is therefore recommended to empty the titration vessel after every other titration and start with a fresh base solvent.

5. Verification of the accuracy and detection limits of this method using standard samples is underway.

REFERENCES

1. D.M. Smith, W.M.D. Bryant, and J. Mitchell, Jr., "Analytical Procedures Employing Karl Fischer Reagent I. Nature of the Reagent", *J.Am.Chem.Soc.*, 61(9), 2407-12(1939).
2. I.M. Kolthoff, and P.J. Elving, Treatise on Analytical Chemistry, Newyork, Interscience Encyclopedia, Part II, Volume 1, Section A, pp 69-165, 1961.
3. J. Mitchell, Jr., and D.M. Smith, Aquametry : Application of the Karl Fischer Reagent to Quantitative Analyses Involving Water, Newyork, Interscience Publishers, 1948.
4. W.M.D. Bryant, J. Mitchell, Jr., and D.M. Smith, "Analytical Procedures Employing Karl Fischer Reagent V. The Determination of Water in the Presence of Carbonyl Compounds", *J.Am.Chem.Soc.*, 62 (12), 3504-5(1940).
5. W.M.D. Bryant, J. Mitchell, Jr., and D.M. Smith, and E.C. Ashby, "Analytical Procedures Employing Karl Fischer Reagent VIII. The Determination of Water of Hydration in Salts", *J.Am.Chem.Soc.*, 63 (11), 2924-27(1941).

     Characterization Techniques for the Solid-Solution Interface

INDEX

     Characterization Techniques for the Solid-Solution Interface